This Smallholding Life
A Delicate Balance

Adrian Doyle

First published in Great Britain in 2020

Copyright © Adrian Doyle 2020

The moral right of the author has been asserted

All rights reserved

No part of this publication may be reproduced, stored in a retrieval system, or transmitted, in any form or by any means, without the prior permission in writing of the publisher, nor be otherwise circulated in any form of binding or cover other than that in which it is published and without a similar condition including this condition being imposed on the subsequent purchaser.

All rights reserved.

ISBN: 978-1-8380010-0-1

Find out more about Adrian and Nicole, their smallholding and shop here:
www.auchenstroan.com

Editorial services by
www.bookeditingservices.co.uk

With thanks to everybody who helped.

CONTENTS

Plans of Smallholdings i

Prologue v

1 Introduction 1

2 Why Choose the Smallholder Life? 5

3 Ivor 14

4 The Search 20

5 Larry and Lisa 34

6 Planning and Balance 39

7 Ymogen 49

8 Smallholding and Money 53

9 Ant and Dec 63

10 Getting Started - Infrastructure 68

11 Bees 74

12 Equipment 79

13 Witchy 84

14 Growing Fruit and Vegetables 88

15 Lucky Thirteen 95

16 Keeping Livestock 99

17 Vera, Vi and Ursi 107

18 Land Management 114

19 Uninvited Badger 119

20 When Friends and Relatives come Visiting 123

21 Borrowing Cows 127

22	The Good Life - Finding the Balance	134
23	Sarka	138
24	Sheep	142
25	Pigs	168
26	Chickens	182
27	Cows	196
28	Bees	209
29	Selene	221
30	Last Word	225

PLANS OF SMALLHOLDINGS

Our Smallholding in Somerset

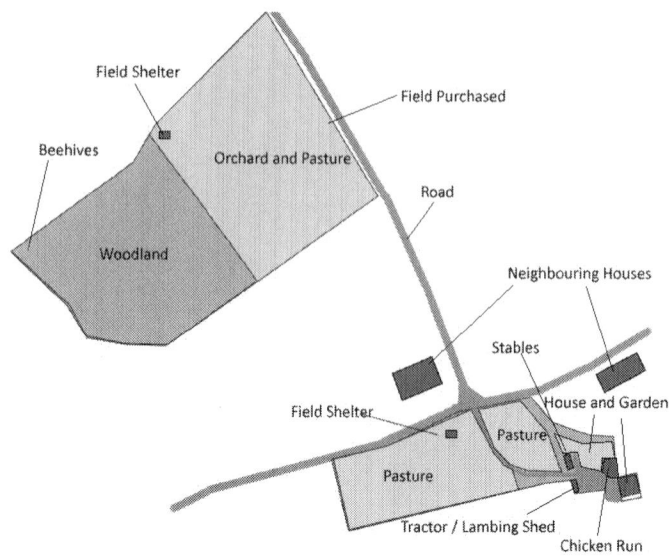

Our Smallholding in South-West Scotland

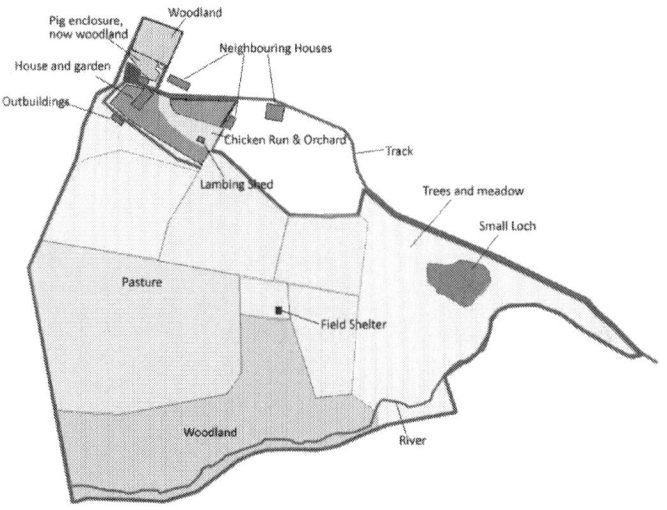

For the latest stories from our smallholding, check out
www.auchenstroan.com

PROLOGUE

Adrian and Nicole have years of experience of setting up and running smallholdings. Over this time, they have kept sheep, chickens, pigs, bees and Highland cows. Exploring ways to make money from their ventures has been a constant challenge in their smallholding life.

Their first smallholding was one and a quarter acres in size and was to be found in Somerset. Soon after moving in, both Adrian and Nicole found themselves working flat out from dawn to dusk. Most of this work was sorting out the land: vegetable plots were dug, chicken runs created and Nicole threw herself into transforming the garden.

There just never seemed to be enough time in the day. Looking for ways to reduce the workload, Adrian and Nicole decided to get some sheep to help keep the grass down. This involved installing new fencing and the purchase of field shelters. However, it was not long before they realised that sheep were high-maintenance animals that required expert knowledge. Furthermore, sheep needed the grass to be cut from time to time. The balance between effort and reward was proving trickier than planned.

More pasture was needed as the flock grew, so Adrian and Nicole bought a field nearby. More fencing was installed splitting this field in two: part woodland, and part orchard and pasture. Over a thousand trees were planted in the woodland section.

Looking back, Adrian and Nicole were run ragged for those first two years.

However, they loved the life and now, with the flock of sheep needing more space, they decided to look for something bigger. Having looked at what was available, they settled on south-west Scotland and bought a much larger smallholding. It came with a newly planted woodland as well as acres of hilly pasture.

The move itself turned into a major logistical exercise. None of the specialist farm removal companies were interested. They could get a removals company to move the house contents, but they were going to have to do the rest themselves. This required gaining certification for moving sheep that distance.

On the move day itself, Adrian and Nicole had a trailer each, Nicole moving the tups and Adrian the 10 ewes. An overnight journey meant light traffic but numerous road closures leading to diversion after diversion including a fraught journey through Birmingham with lorries thundering past on all sides.

Nevertheless, as the sun rose, they arrived and unloaded the sheep into their new home.

Having moved, Adrian and Nicole found themselves once again sorting the place out to make it work. In fact, it took another two years, during which they carried out a number of home improvements, built outbuildings, installed field water systems (to provide drinking water for the sheep) and installed hundreds of metres of new fencing.

Once again, they found themselves being run ragged.

Looking back, they realised that while they love this smallholding life, it is both time consuming and expensive. In fact, they had spent a lot of money in setting up their two smallholdings to work for both themselves and the animals they kept.

And from that experience came this book, a practical insight into both the financial side of smallholding life along with the emotional highs and lows.

1

INTRODUCTION

This book is intended as a guide for those of you thinking about living more self-sufficiently via moving to a smallholding or something similar. The focus of the book is on the practical challenges you will face along with the good, the bad and the ugly aspects of this lifestyle.

There is much written material already available on this subject. You can read up on sustainable living, the agony and ecstasy of smallholder life, and there are even manuals on how to live off-grid. All of these are an important contribution. So, you may be wondering, why write another such book? Well, one thing that I feel would really help is a book that provides an insight into the economics of such a lifestyle. This is what this book seeks to address.

There are lots of challenges to this way of life. Keeping animals, keeping bees and growing your own food all present their own unique challenges. But, for me, the biggest challenge is getting the balance right between time, effort and money.

It is a fact that there are many ways in which urban and rural living overlap. One of these is the constant array of jobs that need to be done. Whether this is cutting the grass, painting the kitchen or feeding the animals, these jobs all have one thing in common: they take time. In urban living and assuming you have a full-time job on "normal" hours, your free time is limited to evenings and weekends. The choice you face is whether to spend your free time doing these jobs and therefore save money, or pay someone to do them and so save time.

The difference in smallholder living is that the number of jobs is larger. If you have ever run an allotment, you will know what I mean. Finding the time to keep on top of it while working full-time presents something of a challenge.

Running a smallholding takes a lot of time. The more you plan to do, the more time it takes. Growing vegetables, caring for livestock, even harvesting and cleaning vegetables all take time. Historically, vegetables came from greengrocers and needed to be washed and prepared. These days supermarkets have sanitised everything, so we have become used to buying produce that we can put straight into the pot. This saves a lot of time as you will find out when you harvest your carrots and find that an army of slugs have got there first.

In living a smallholding life, the key point is that there is a lot to do, so it is hard to keep up if you are in full-time employment. As this book will demonstrate, it is also very difficult to earn a living from a smallholding. Prices of food are as low as they have ever been, so this limits how much you can sell your produce for. Supermarkets pushing prices down has led to the consolidation of farms into large agribusinesses. These large-scale farms have costs lower than you can achieve – you can't compete, not that you would be wanting to sell to a supermarket anyway. The point is, people who might be your customers have got used to supermarket prices and supermarket convenience.

While there is a market for high-quality produce, it is not as big as you might think; and much of it has been cornered by specialist farm shops that are now to be found everywhere.

This is important because while you may be able to produce most of the food you eat, there will always be bills to pay. This is definitely the case in the UK, and also true of other countries. So, smallholders need to generate an income.

Throughout the book, I use the council tax as an example of a bill you cannot avoid. Wherever you live, the council will require you to pay this tax. There's no escape. In reality, that is just one bill of many as there are a range of outgoings you still need pay in the country including, for example, transport, heating, internet, phone, and so on. But, for the sake of simplicity, this book shows examples of how much work is needed just to generate enough income to pay the council tax.

What this book also explores is how to find the balance between the conflicting demands of time, money and effort. It explores the choices you will face and the possible impacts on each of the above. With a little forethought and planning, you can make the transition to smallholder life go a lot smoother and maybe cheaper too.

I will also tell you about the highs, lows and stresses of smallholding life. If you have ever kept animals, you will know that they need to be looked after every day. While you can find people and businesses that will look after your cat or dog while you are away, this is much harder when you expand your animals to include, for example, chickens, sheep or even cows. In fact, if you keep livestock, you are moving into a world of working 7 days a week, 365 days a year.

You are moving into a world that is much closer to the natural cycle of life and death. If you have had a pet pass away, you will have some idea of what I mean. If you keep farm animals, you will need to prepare yourself for the fact that they die, and not always when you might expect them to.

The highs and lows of smallholder life are much more pronounced. It is much more of a rollercoaster type of living.

Is this type of life for you? Having grown up in a city

myself and worked in corporate life before moving to the country, all I would say is that I can't recommend smallholder living highly enough. It is an existence that is more authentic than anything else I have experienced.

Some people like the thrill of rollercoasters because, for a few seconds, it makes them feel more alive. You get highs and lows on a daily basis from running a smallholding.

2

WHY CHOOSE THE SMALLHOLDER LIFE?

There are many benefits to smallholder life – the obvious one is that you'll be living in the country. This book is mainly focused on helping those who want to live off the land, to some extent. It is what is commonly called "the good life" and, indeed, it can be.

One of the best reasons for living the "good life" is your health. There are many ways in which your health can benefit, and some of these are outlined in this chapter. But, basically, the real improvements are better food, better air quality and a more active lifestyle.

Where do we start? We humans need a few basics in order to survive. These are air, water and food. Air is an interesting place to start. At the time of writing, the UK government has been repeatedly taken to court by clean air campaigners. Basically, the UK government stands accused of doing little to improve air quality across the UK, especially in cities.

Pollution has been with us for decades, if not centuries. It is all down to burning fossil fuels. Historically, people heated their homes with open fires that burnt wood, coal or peat. As cities grew more populated, more smoke was generated. London was infamous for its smog.

In the twentieth century, central heating started to take over and the smoke reduced. At the same time, the car replaced the horse. These days, most of the pollution in a city comes from cars. It may not look like the legendary London smog, but it is a problem, nonetheless. The thing is, these toxic particles can be too small to see (unlike smoke-fuelled smog), so they are much easier to ignore.

In the countryside, there are less people and less vehicles. The air is naturally cleaner. Not only does this mean we are breathing cleaner air, but it is also good for your skin. Just think about blowing your nose after a trip on the London underground. All that black, sooty matter also settles on your skin. So even just moving out to the country has health benefits.

To some extent, the same could be said for water. Many country properties have their own water supplies. Mains water in the UK is treated with chlorine to make it safe to drink. It is also treated with fluoride to help teeth. If you have your own water supply, you will have less chemicals in your water. I am not saying mains water is not safe to drink, far from it. I just prefer the taste of water without chlorine. And given the huge market for water filters and bottled water, I suspect I am not alone.

The rest of this chapter assumes that you will be producing some of your own food. Moving to the country, in itself, doesn't mean better food. Supermarkets are everywhere. Farm shops selling organic, locally grown produce can be found in both town and country. So we all have access to quality food if we want it.

However, there is a great deal of satisfaction in eating something you have grown yourself. This is even more pronounced when you have grown the food from seed.

Since the middle of the twentieth century, food production has moved away from traditional small-scale farms to large-scale agribusiness. Traditional farming involved rotation of both crops and livestock. What the plants took from the soil, the animals put back.

Over the years, both the production of meat and

vegetables has become big business. At the same time, supermarkets have been driving down prices. They have also been getting ever more selective about presentation. Much edible food is discarded simply because it is the wrong shape.

This has, in part, led to what we call factory farming. This applies to both animals and crops. Both approaches are harmful to our planet. There are many sources of information about climate change and its relation to the food we eat, so I won't go into too much detail here. Information on factory farming is also widely available. For animals (meat and dairy), factory farming means packing animals into ever denser conditions. Ironically, legislation is forcing egg and chicken farms to give more space to hens. But, at the same time, we are packing pigs and cows into ever smaller spaces.

This high intensity means that there is no natural foraging available for the animals. They are fed grain, which, in turn, must be grown. So, if you think about it, it is still using a lot of land, but the animals don't have access to it. And whatever modern factory farmers might tell you, I cannot see how cramped conditions, a lack of sunlight and a monotonous diet is good for any creature.

There is a rising groundswell of opinion pushing back against factory farming of animals. The two most visible drivers seem to be climate change and the rise in veganism. I can see, and to some extent agree with, the arguments put forward by these movements. However, I also feel it is important not to ignore the damage caused by the factory farming of crops. The water crisis in California caused by large-scale almond farming springs immediately to mind. But closer to home the key word is "soil". Intensive monoculture irreparably damages soil. If it continues, there may come a day when we won't be able to grow anything.

Large-scale crop farming uses chemicals. Fields are sprayed with weedkiller to kill weeds. If you want to investigate this, look up "glyphosate", a weedkiller commonly used worldwide until recently. You can add to that the extensive use of pesticides such as neonicotinoids, for example. These are alleged to harm bees; a contention against which the chemical giants are arguing vociferously. Without bees, we all die. Einstein himself predicted that without bees, humanity would survive no more than four years. We need bees to pollinate our plants. Yet, agribusiness appears to prefer to gamble our future by marketing these pesticides to our farmers for profit.

Residues of these chemicals can be found on the fruit and vegetables available in our shops. Some are even inside the plant, having been absorbed. Washing your fruit and vegetables, even though they look ever so clean in their custom plastic wrapping, doesn't necessarily make the food clean.

Factory crop farming also wreaks destruction on wildlife insofar as it creates large sterile areas.

By growing your own, you can ensure chemical-free organic produce.

I haven't even got to taste yet. I remember picking strawberries and filling punnets as a child. I love strawberries. Yet now, I find the strawberries from the supermarket are often tasteless and bland. Even in season, sourced locally, I think they have lost much of their flavour, especially compared with what we grow on our smallholding. But more on taste later.

A lot of the above also applies to meat. Intensive farming can harm animals both physically and psychologically. Under these conditions, the animals can become less disease resistant. It is a bit like a human couch potato: little exercise and lots of food. It is not good for us, and it is not good for animals. Faced with these challenges, what do large-scale farmers do? Encouraged once again by agribusiness, intensively farmed animals are given routine doses of antibiotics. Traces of these can be found in the meat you buy from supermarkets. This widespread use of antibiotics in farming has been cited as a major cause of large-scale antibiotic resistance in human medicines. For example, methicillin-resistant staphylococcus aureus (MRSA) is a bacterium that causes infections in different parts of the body. It is tougher to treat than most strains of staphylococcus aureus – or staph – because it is resistant to the most commonly used antibiotics.

The net result of this is that many of the medical advances made in the twentieth century could be negated by our inability, as humans, to fight off infection.

All the above also affects the quality of the meat, especially the taste. Like the strawberries, factory-farmed meat, to me, has lost much of its flavour.

What has this got to do with smallholding life? Simple, you are getting away from all of that. A real benefit is that you can control what you eat. Aside from being tastier, your own produce is probably much healthier. It still has all the vitamins, proteins and minerals that can be lost in processed food.

Also, by eating your own produce, you are having a small but real effect on climate change. Less food miles for a start.

Regarding meat, you can, if you so choose, produce your own meat. If you do, you will be surprised at the difference in both quality and taste. An animal that has lived a life as natural as possible will be a happy animal. That comes through in the meat.

Aside from growing vegetables and rearing livestock for meat, you can also keep chickens for eggs and bees for honey.

Home-produced free-range eggs taste so much better than eggs bought from, for example, the supermarket. You really have to taste them to see the difference.

Bees are slightly more specialist. We have experience of keeping bees, and it can be great fun.

Suffice to say that whatever food you produce, you will be amazed at how much better it is. It might take more effort, especially when it comes to preparing it, but it is worth it.

Back on to health. We have already looked at the benefits of clean air and home-grown food, but there are other benefits to be enjoyed in smallholder life.

For us, the top of the list is the escape from the "rat race". Before moving to the country, we both had corporate lives. The five-day week at the office, and the evenings and weekends blighted by emails and mobile phones. The scrabbling to the gym to try and keep fit. The lunchtime sandwich grabbed at the desk. Busy, busy, busy. We had an allotment but struggled to get there more than once every two weeks or so. It was not that productive!

For some, corporate existence is what they live for. It works for them. For others, like us, it was a job, but not really a way of life. And for us, our expenses always seemed to eat up all of our pay. It was a hard and unrewarding treadmill.

It is possible to retain full-time jobs and at the same time run a smallholding, but that can be a lot of work. We have met a few couples where one partner keeps their job and the other manages the smallholding full-time. It is really a personal choice. For us, the ideal is part-time work that allows us to keep up with all of the farm tasks.

You might be wondering why you need to work at all. The bottom line is that it is hard to generate a living exclusively from a smallholding. Most people living this life have extra jobs. The Highland cow farm up the road is also a haulage business and hay supplier. Another smallholder up the road works as a freelance car mechanic. Down the road, a smallholder works as a painter and decorator and also does a bit of tree surgery.

Part of the reason for this book is to highlight the need to be able to balance the finances and small-scale farming.

Later chapters provide a lot more detail on this.

Smallholding life still has plenty of its own stresses, but they are different. Gone are the endless deadlines and high pressure. What comes in is hard work and a whole new set of worries.

The hard work really depends on what you choose to do. A basic rule of thumb is that the more you choose to do, the more work you will have. Plus, everything has its seasons. Growing vegetables becomes a hectic activity in the spring as you plant out seed trays, pot up, plant out, and so on. Blink, and your carefully weeded and mulched patch can be covered in weeds.

Keeping animals takes time, too. As a rough estimate, I would say that each type of animal requires around an hour a day.

So, returning to the idea of work-life balance mentioned throughout this book, if you are planning to keep a full-time job and run a smallholding with animals, you will need to plan your time carefully.

Animals get into all sorts of trouble. They get ill. They get trapped. They escape. If it can happen, it will. The point is, it all ends better if you are around to catch it early. For example, sheep can get stuck on their backs. It is not a good position for them to be in as their internal organs settle in an uncomfortable way and they become targets for crows and ravens. If they are left like this for long, they will die. Catch them in the first hour or so and they will get up and walk off none the worse for wear. The longer they remain prone, the more their chances of recovery diminish.

Returning to the subject of stresses, one of the smallholding stresses is looking after animals. They keep you on your toes. The flip side, though, is that keeping animals is hugely rewarding. We are a nation of pet lovers. Keeping livestock can be an extension of that. All our animals have names. Each has its own personality. We know each animal individually. And the sheep know us too.

And yes, even sheep have their preferences as to which humans they like best.

What is noticeable is that, with animals, you become closer to the natural cycle of life and death. Animals die. And they do so for all sorts of reasons. And it is hard when they do. In this smallholding life, the emotional peaks and troughs are much greater than other lifestyles.

Nothing highlights this more than breeding. The intense joy at helping a ewe deliver a lamb for your first time cannot be matched. The deep satisfaction of watching a sheep go from a nervous and jumpy creature to a contented first-time mother is amazing to witness. Having lambs use you as a toy to play on. They can bring you so much joy. Yet, after delivering a lamb, having another die in your arms is heartbreaking, more than I can describe.

Yet, this is a truly authentic existence. Because, deep down, you know you are doing your best for these animals. You are making a difference.

Keeping animals is hugely rewarding. I could go on and on. The sense of achievement when scratching a Highland cow, or bull even, behind the ears. Watching chicks clambering over your hands pecking at crumbs. Scratching a pig's tummy because it has rolled over hoping you will do just that. Calling a sheep by its name and watching it run towards you and into the pen while the vet is standing next to you with a surprised look. Smallholder life is full of such moments.

If that's not enough, there are other benefits.

One of these is that you learn new skills, "proper" skills, which makes them even more satisfying. For example, through necessity I have become reasonably accomplished at building and repairing stone dykes. I have also learnt how to put up fencing, construct outbuildings and carry out agricultural plumbing. Nicole has become an expert at making felted woollen rugs, animal husbandry (including giving injections), sheep and cow calling, helping lambs come into the world and diagnosing sheep ailments. We can also both shear sheep. These new skills, aside from being satisfying in their own right, contribute to a deeper inner confidence. If you can handle the things that this life throws at you, you can handle just about anything. It is a good feeling.

In a way, smallholding life is a return to what we humans were designed to do. It certainly feels a lot more natural than sitting in an air-conditioned office staring at spreadsheets.

The huge plus points are that the active lifestyle, the outdoor living and the good food all make for a more balanced, healthy and rewarding lifestyle.

3

IVOR

We love Highland cows. There's something majestic, almost magical about them. This love affair with Highland cows was strengthened when we visited a neighbouring farm that bred them. We met the bull, Hamish, a gigantic beast who stood just the other side of a one-metre high wire stock fence. He dwarfed both us and the fence. His horns were so wide you could not reach across them with your arms fully extended.

Hamish looked us up and down and then turned sideways on proffering his back for a scratch.

Our grass was getting quite long in places, and sheep like it fairly short. They like to nibble rather than pull it up in clumps. Cows like it long. Also, having cows share the pasture with sheep reduces the sheep's worm count.

It was a no-brainer. We bought two, Bluebell and Texa. It was not quite as spontaneous as it sounds. We did check what their needs would be, and were assured that all they needed was grass, water and a bit of training. They are hardy beasts and survive happily in near-Arctic conditions. South-west Scotland is positively balmy for them. They'll be no problem at all, we were told.

We didn't take them home with us straight away. The fact that we didn't have a cow-sized trailer was one factor. Also, the sellers told us it would be a good idea to get to know them a bit before they moved. Sounded like a plan. Building trust with animals, especially herbivores, takes time. These are prey animals, and humans are perceived as predators. They are wary of us by instinct. So, we made regular visits and learnt how to approach them and scratch their backs.

Then it was discovered that Bluebell was expecting. This was unplanned. Seemed like Hamish had made a secret visit when no one was looking. Hamish and Bluebell are related, so the calf would be inbred. We didn't mind as we had no intention of breeding. We were just excited at the prospect of having an extra Highland cow.

This delayed delivery as it was thought best by all of us to allow Bluebell to give birth on familiar territory. Moving her before birth might have caused her a lot of stress. It is a common theme we have found with animals. We feel we are bringing them to a fabulous new home. But the animals just see themselves as being torn away from what they know. So, it all needs careful handling. And time.

Anyway, a few weeks later, the three Highlands arrived. The calf was so cute. He was a boy, and we called him Ivor.

Ivor initially kept his distance. Although his mother was Bluebell, it was Texa who took on the guard duty. If we got too close, Texa was straight in. She'd stand there and give us a look that pretty much said "be very, very careful". When you have two big eyes and two huge horns pointing right at you, you are very, very careful.

Nevertheless, each day, Ivor got more curious. In between skipping around the field, he'd wander over close and inspect us. One day, I was sitting there watching the cows when Ivor wandered towards me as he had done, tentatively, before. Only that day, he didn't stop. He came right up and touched my nose with his nose. I could hear Texa snorting behind me, so I kept very still. I think Texa wasn't quite sure what to do. By now, we had established a relationship with Texa and Bluebell and shown ourselves to be no threat. However, Ivor was a bit too close to a human for her comfort. It wasn't long before Ivor grew bored of inspecting my nose and wandered off. I rushed off to tell Nicole what had happened.

Jealousy is a funny thing. Nicole was both delighted and aghast at the same time. She wanted her nose to be touched by Ivor.

She didn't have to wait long. On her next visit, Ivor wandered up and licked her face. Licked her face! I was most miffed. And so it went on. With each visit, Ivor grew more and more curious and spent more time checking us out. It was not long before we were able to scratch him behind the ears.

The thing with Highlands is that they love their scratches, so long as they are from the neck backwards. Touch the horns or the head, and you can find yourself bracketed by two horns and the subject of a steely stare. It can be done, but it takes a long time. While Ivor was letting us scratch him from head to toe, Bluebell and Texa had just about let us get as far as their shoulders. The best approach with Highlands is to start at the back and work slowly forwards. If they shake their horns, move slightly backwards.

It is fair to say that we were besotted with Ivor. It took a lot of time out of our day, but it was worth every minute.

Winter came early that year, and by November the first blanket of snow had covered our fields. We had made hay that year, so we started feeding it to the cows. They loved it. They loved it so much that they were suddenly going through two small bales a day between them. It didn't need a degree in mathematics to work out we were going to run out very quickly.

That started the mad dash to source hay. It turned out there was not a lot about that winter. I found a small farm that had about 100 bales for sale. We took the lot. That was still not enough. We needed to think this through.

In the absence of hay, we looked to get haylage or even silage. Haylage and silage usually come in large round bales and weigh hundreds of kilograms. They are the big round bales you often see in fields or stacked in farmyards. The trouble was, we had no equipment to handle them. So it was emergency shopping time. Fortunately, we have a compact tractor, and I was able to source a spike attachment so we could use the tractor to move them.

Then we had to find a haylage or silage supplier. We asked around and finally found someone with a few bales of haylage left. Then it was "scrounge a trailer" time. We had two trailers, but neither would be capable of carrying haylage bales. Finally, it was off to collect them along the narrow, twisty lanes of rural Scotland which were, at that time, covered in ice.

I brought some bales back and parked up. But could I unload them? I could hardly move them, let alone roll them off a trailer. In the end, I had to wrap a strap around them, tie it to a nearby tree and drive forward. That did the trick.

We had previously acquired a large bale cattle feeder, which is a circular metal framework that fits nicely around a large round bale. We'd been putting small bales in – it stopped the cows from trampling the hay.

We attached the tractor bale spike and took the first bale down. We stripped the plastic and netting off the bale and then tried to tip it onto its base. A few minutes later, having managed to move it at least two inches, we stopped to ponder. I then headed off to bring back the tractor. That made life much easier. With a gentle nudge from the front loader, the bale was in place in no time.

The new feeding station was in a completely different place. Given that we were now using the tractor, it had to be easily accessible. Up until now, we'd been feeding the cows hay in the middle of a field halfway up the hill. The new feeding station was next to the track near the bottom of the hill.

So, we rolled the feeder down the hill and over bumpy, frozen ground and icy sheets. At least it was downhill!

With everything finally in place, it was time for the best bit: bringing the Highlands down to their new food. They had, of course, migrated to the top of the top field, as far away as possible. They do like it up there. Nevertheless, armed with samples of haylage to use as bait, we soon had their attention and they started to follow us down.

Now, Bluebell and Ivor were calm as you like. But Texa could get very excited and do an excellent impression of a charging, bucking bronco. She usually skidded to a halt beside us looking slightly surprised and also gazing intently at any food we might be carrying. In an effort to keep her calm, I dropped a little haylage on the ground in front of her. Luckily, this kept her happy, and we kept going with the others. After that was eaten, Texa bucked and charged again, causing Nicole to drop her bundle of haylage. Texa didn't mean to hurt us, she was just excited and happy, but when you are standing on an icy hill watching her careering towards you, you do worry that she might misjudge slightly!

We soon had Texa down at the feeding station. Bluebell and Ivor had stopped on the way having found some tasty grass, so off I went to fetch them. I persuaded Bluebell to follow me without too much trouble. She loves haylage. Then we heard a plaintive "moo" from behind the wall. Ivor had stayed behind and was wondering where everyone had gone. Nicole went off and fetched him.

The cows loved their haylage. They got so excited at refill time that we had to lock them in another field as we manoeuvred it into place. And it got harder. The cows had churned up the feeding area, so it was touch and go each time whether the tractor would slide off the track and down the bank. As the tractor driver, my heart was in my mouth each time.

It was a pretty tough winter. At one point the snow was so deep it was coming in over the tops of our wellies. It was actually easier when it was cold and the mud frozen. Trudging through deep snow is pretty tiring. Trudging through deep snow and mud is exhausting.

It was that winter of 2017 when Britain was hit by a weather front named "The Beast from the East". When "The Beast" hit us it was cold, but we did not get that much snow. It all fell on the east coast for once. We had a few snowdrifts, but the roads and tracks remained passable. Well, we have 4x4s, so they were passable to us. We did, as it happens, find our neighbour stranded by the side of the road one evening having abandoned his van. I am pretty sure he appreciated the lift home.

Our morning routine is that Nicole carries out the animal checks and I prepare breakfast. I like to think that being a Scot, one reason for this is that I make better porage. We are both up early, but we each have one day a week where one of us has a lie-in while the other does everything. Saturday is my lie-in day.

One snowy Saturday morning, Nicole came running in, in a state of panic. "Ivor is dead!" she screamed. Within seconds, I was fully clothed and running alongside her through the snow. Ivor had collapsed next to the feeder. Texa and Bluebell were in close attendance. As I checked him, I saw that he was breathing, just. He was still alive. We stripped off all our protective clothing and covered him. Nicole ran off to get quilts and hot-water bottles and to call the vet.

We covered Ivor and did everything we could to warm him. However, the vet looked worried. She said the prognosis was not good. She did her best to warn us, but we couldn't give up. By now, some of our neighbours had arrived. I said we needed to get him off the snow and indoors. We have an Aga range cooker – if we could get him in front of that, there might be chance. I ran off to get the tractor (there was no way we could carry him) and between us, we managed to get him all the way to the Aga.

It was a marathon feat, but we managed it. We laid Ivor next to the Aga in the warmth of the kitchen, but Ivor had had enough of this world and died in our arms.

4

THE SEARCH

Once you have decided to go for the smallholding life, you need to find somewhere to move to. With modern technology, in other words the internet, this is much easier than it used to be. As well as the "standard" property websites, there are both specialist sites for farms and smallholdings as well as websites focused on particular areas that are worth keeping an eye on. These sites cover both properties for sale and those available for rent.

Visiting properties can be turned into a fun experience. When we were looking, we grouped possible properties together and arranged viewings over a couple of days. Then we'd find a nice B&B with a pub nearby that sold good food and make a weekend break out of it.

It is not a bad idea to stay over in the area you are looking at. Aside from the fact that these properties could be some distance away, you can also see what the surrounding areas are like. A move to the country is not just a case of moving house. You are moving to a new lifestyle, so you'll be surrounded by people who are likely to be very different from your previous neighbours. Don't worry, the difference is a good thing. Most country dwellers have some links to farming and other countryside activities such as horse riding. They just have different priorities to city dwellers. So, it is good to immerse yourself in it even for a day or so just to get a sense of the place that you are thinking of moving to.

And all this brings us to some of the decisions that will be facing you. The first of which is likely to be where do you want to live?

This decision will be affected by many factors: childhood memories, where you grew up and what you have seen on TV, to name a few. The first thing is to make the big calls – such as are you planning to stay in the UK or are you thinking of moving abroad? There have been some interesting programmes on TV about people and families who have chosen to move overseas for a life in the wild. I have watched programmes in which people have relocated to places as far apart as the side of a volcano in Chile to the frozen tundra of Northern Canada and Alaska. Some move into the forests in countries such as America, Panama and Belize; some to Eastern Europe; some to tropical Asia; and some to Australia, New Zealand and even the Pacific Islands. Others try to eke an existence in sub-Saharan North Africa. Some only move as far as France or Spain. Each of these destinations has a wide range of factors to consider. A lot of it will come down to your personal preferences. This book, however, focuses primarily on the UK.

Even within the UK, there are a wide range of destinations – each very different. If we start in the north, Scotland has many islands. These include the Shetlands, the Orkneys and the Hebrides. Historically, these areas were inhabited by crofters, so there are many crofts and small farms here. If you like stormy, wet weather, then the islands could be tempting. Most have coastlines on the Atlantic Ocean, so there is little protection from the incoming weather systems. While this can be fine when you are tucked up in front of a warm fire, it can play havoc with outbuildings, especially polytunnels.

Northern Scotland is fairly mountainous. Not all breeds of sheep are adapted to this environment. So, if you are thinking of keeping livestock in areas such as this, they will need to be a hardy breed. It is also difficult, if not impossible, to grow arable crops in the mountains.

As you move south, the climate becomes gentler. In general terms, the west coast is wetter than the east coast. Also, as you drift south, the average temperature rises. It is around a four degrees (centigrade) average increase from northern Scotland to southern England.

Continuing southwards, the mountains give way to flatter land around Tayside and the central belt of Scotland (the triangle between Glasgow, Edinburgh and Stirling). The central belt is quite built up with several towns and villages close to each other, but there are still plenty of farms in this area.

The Borders and Galloway areas of south Scotland are more orientated towards agriculture, so there is more scope to find places here.

In northern England, you can choose between the Lake District, Lancashire, Yorkshire and Humberside. Each has its own unique charms.

There is plenty of choice in middle England, and this choice expands into the southern areas of England. It can be quite expensive in and around the Home Counties, but as you move further away from London, you are likely to get more for your money.

Looking west, Cornwall and Devon are quite popular destinations for smallholder living, as many of the books I have read attest to.

And there is also Wales and Northern Ireland. Wales is quite a mountainous region, much like parts of Scotland, only a bit warmer. Wales has something of a reputation for having lots of rain, but this is not necessarily a bad thing for smallholder living.

Having lived in both Scotland and England, I can tell you from personal experience that it definitely feels warmer in the south.

Wherever you are thinking of, it is worth taking the time to visit. When we were considering moving from Somerset, we were thinking about Scotland or even moving abroad to somewhere like Switzerland (my mother-in-law is Swiss, so we know Switzerland quite well). We booked a last-minute trip to Scotland, to Stirling to be exact. We spent three days looking around, from Tayside to Loch Lomond and all the surrounding areas. We spent time in local pubs (in Bridge of Allan), chatted to local people and I introduced Nicole to the pleasures of the white pudding supper (which you can only get in Scotland). We loved it, so we started looking at Scotland as a destination, and here we ended up.

Within all of these areas, there are some core decisions to think about. These include things like how remote you want to be, how much land you want, what sort of house and outbuildings you want, and what you want to achieve. Let's look at some of these.

Let's start with thinking about how remote you want to be. Some smallholdings and small farms are very remote with few, if any, neighbours. On the other hand, some properties are close to towns and even cities. And there are plenty in between. This really is a personal choice. I was city born and raised (Edinburgh), as was Nicole (Slough). Yet, we live in a fairly remote area of south-west Scotland. There are approximately eight houses within walking distance and the nearest town is four miles away. This suits us fine, but it is a far cry from what we were used to. Just up the road, the properties are much more remote with 2-3 km between each house. The point is, it can take a lot of thought and research to work out what you really want.

For example, when we first started looking at a move to the country, one of our important conditions was that we had a pub within walking distance. This was not because we wanted a good drinking hole, it was because we wanted somewhere we could go to eat, have a glass of something and then walk home. Preferably, a place that welcomed dogs, too.

Living a remote lifestyle can also be something of a trap. It can become quite easy just to stay at home every day and carry out the endless list of tasks that need doing.

That said, we make the effort to go to the pub in town at least once a week, and we both head into Edinburgh on a regular basis. We need our fix of civilisation! Our local town benefits from two pubs, an excellent restaurant, a great cafe and a decent shop. It all helps.

So, living out in the sticks or bordering a town or village. It is something to think about.

This also ties in with another thing to consider. How often do you want friends and family to visit? The more remote you are, the less frequently this is likely to be. Unless, of course, you live next to something of great interest like a castle or an amazing beach. Having friends and family to visit is quite an interesting topic. We have found that many of our friends and family have a very different idea of smallholder life from its realities. We sometimes get the impression that they think we are on one giant holiday, sitting on the front lawn in our deckchairs reading books all day. The reality is entirely different, although this depends very much on what you choose to do. We have sheep, and they are high-maintenance animals, so they can take a lot of time. We are busy all the time.

Nevertheless, it is another thing to consider. We are about an hour and a half from two major airports and an hour from a mainline train station, so our friends and family have options when they come to visit. At one point, we had been considering a property to the south of the Mull of Kintyre. It is probably a three- or four-hour car journey to Glasgow from there, so something of a long round trip to pick up someone from the airport. It was just too far away for us.

What kind of house would you like? This is not as simple as it sounds. Sometimes, the location and land available can make the house seem a secondary concern. But it is worth thinking about. Different parts of the country are likely to have very different choices. In Scotland, many rural properties are old farm cottages. Many have been extended. Many are old and poorly insulated. But they all have character!

Other parts of the country will have different styles. In the south-west, there were many barn conversions available. So, there is a choice out there. You can get a feel for this online through the websites with houses for sale.

In our experience, you can find anything from a small cottage to a stately home. It all depends on your needs, and your budget.

This ties in with how much land you want. And this will be driven by what you want to do. Or what you think you want to do. I say this because things change. Our first place, in Somerset, came with around one and a half acres. With Nicole coming from a modern flat and myself from a modern housing estate, it seemed vast! Yet, within three years, we had bought another field increasing our acreage to five acres, but that was still not enough! This is one of the reasons we now live in south-west Scotland.

The two main factors that drives how much land you need are crops and animals. Vegetable patches need not be too big, unless you are considering commercial growing of vegetables. However, we have never met anyone, yet, who moved to the country simply to grow vegetables commercially. To be honest, it is hard enough growing them just for your own consumption.

Regarding animals, the assumption is that you want the animals to live as naturally as possible, so for sheep and cows this means a diet of grass. We consider some of these animals (and more) later in the book.

There are many different recommended numbers of animals per acre on the internet. Much depends on the quality of the grass. Based on our experience, as a rule of thumb, you will need around a third of an acre per sheep and/or around two acres of pasture for each cow. Both are social animals, so you will need at least two cows and/or two sheep. It is better to have larger flocks of sheep, though, as the more there are, the more settled they are.

You will also need winter quarters for most breeds of cows. There is more detail on that in the chapter about cows.

So, assuming you wanted 12 sheep and 2 cows, you would need 4 acres for the sheep and 4 acres for the cows. That's a minimum of 8 acres, just for summer grazing. Assuming you would also like some hens and maybe a decent-sized vegetable patch, a garden and an area for outbuildings, in this case you would be setting a minimum of 9 acres (allowing an extra acre for the house, garden, etc.). If you are planning to do any breeding, you will need even more land.

Another factor regarding animals is that the number you need will depend on whether you are hoping to generate an income from them.

Still on the subject of animals, it is worthwhile checking out the fencing and outbuildings. If the fences all look old and decrepit, you will soon be facing a large bill, not to mention a lot of work if you choose to do it yourself. Some costs of fencing are included in a later chapter to help you budget for this.

The outbuildings are key if you are planning to keep animals. The primary use for outbuildings is storing feed and housing animals in the winter. Outbuildings are also often used to help with lambing and calving. It can make things easier to have the animals indoors when giving birth, especially if you need to call the vet.

Later chapters in this book go into detail about the kinds of infrastructure you need for keeping different types of animal. This includes details about buildings, fencing, pasture and also handling areas.

At this point, these are just things to bear in mind. The more that there is already in place, the less expense and work you will face when moving in.

To give you an idea, in both the smallholdings we have lived in, it has taken around two years and thousands of pounds to get them set up properly. It is not cheap!

Another thing that is worth checking is where your local farm vet is located. This may not be the nearest vet; many vets these days only treat small animals such as cats and dogs and pass the farm calls on. You need to know how long it will take for a vet to get to you.

Treating animals is different with livestock. While every treatment or inoculation for your dog requires a visit to the vet, with livestock you can do a lot of it yourselves. This includes administering drenches (for example for liver fluke) and injections. Most animals need annual inoculations, and these you can also do yourself. Some vets are happy to show you what to do. This does not, however, mean that you can be too far from a vet.

When you do need a vet, it is most likely to be an emergency, so you will be wanting them there as soon as possible.

One of the times you are most likely to be calling the vet is when a birth is going wrong. Another is when you find an animal that has collapsed. With sheep, for example, all sorts of things can go wrong at lambing. The lambs can get tangled up, they can present the wrong way round (breach), and so on. When this happens, it is that emergency call to the vet. Believe me, a 30-minute wait when a lamb and ewe are in peril feels like an eternity. Every second feels like a minute. Every minute feels like an hour. Ironically, here in the hills of south-west Scotland, we have a much better service than we had when we lived in Somerset. In Somerset, all of the vets in the two nearest towns to us were only interested in pets. Our livestock vets had to come from Devon, over an hour away. Therefore, wherever you are, it is really worth checking this.

Back to animals. They need to be fed during the winter. This means you will need winter feed. You can choose to cut your own hay, or you can buy it in. Either way, you will need somewhere to store it. It is much better to buy it in bulk rather than be nipping down to the local agricultural store and buying it a bale at a time. The winter feed period will depend on where you choose to live. Winter lasts longer the further north you go. And by winter lasting longer, what I really mean is that the grass-growing season is shorter the further north you go. Once the grass stops growing, you need to set up feeding stations. Here, in south-west Scotland, the grass stops growing in late October and starts again in late April. That's about six months with no grass. So, a large shed can be helpful. That said, large bale haylage and silage often comes individually wrapped in plastic. You can see piles of these dotted around farmland. These can be stored outside, so no shed needed.

However, once open, haylage and silage generally lasts about a week. That's fine if you have lots of sheep or cows. If not, you will need hay, which is generally not wrapped but must be kept dry.

That's most of the main aspects to consider regarding animals. I'll cover water a bit later. In fact, all of the above concepts are covered in more detail later in the book.

I mentioned arable crops. I have not met any smallholders who grow arable crops, but it is a possibility. You would need some specialised equipment. Forget those romantic pictures of oxen-drawn ploughs and scythes! It is hard work – mechanisation really helps. So, if you are thinking of growing wheat or barley or similar, factor in how many acres you will need each year, then multiply that by three. Why three? You need to allow for crop rotation.

Crop rotation has been fundamental to small-scale farming for generations. It is part of the wider subject of land management which is covered in its own chapter later in the book. It is not really something to worry about too much at this stage aside from ensuring that you have enough land to allow for this.

To be honest, you can work all that out once you are in and have your animals. But here are a few key points to think about.

Water. Water can be loosely divided into three categories: drinking, flooding and drought. When we were looking at our current place in Scotland, I asked about water in the fields. The previous owners kept sheep, so we were hoping there would be drinking water in the fields. We were told that they relied on streams. The property is bordered on two sides by rivers, and there are, indeed, a few streams. But, on moving in, we found that water was only available in one spot and that it was very marshy.

Marsh and sheep don't mix. And even if they did, you can't rotate the animals through different fields if the water is only in one place.

Two rainwater tanks, two underground water storage tanks, long lengths of pipework and numerous troughs later, we now have water in all the fields.

This also extends to vegetables. Even here, in wet and windy south-west Scotland, we have periods where we need to water plants. So, find out about the water situation.

Flooding. In Somerset, we lived in an area referred to as The Levels. Our property didn't flood, but once a puddle had formed in the winter months – it was there until May. Even with three sheep per acre, or two cows over four acres, it can soon be muddy. And I mean very muddy. The sort that sucks your boots off.

So, look at the surrounding area. If you are in a valley with flat pasture either side of a river, you are probably looking at a floodplain. That can be fine if you also have access to higher ground. If not, you need to ask yourself where the animals will go when it rains heavily. Animals can cope with wind, rain and snow. However, like us humans, relentless exposure to muddy conditions saps their strength.

The moral of the story is think about water and where it will gather during wet periods. We get plenty of rain here and it gets muddy. But we are on a hill, so plenty of the pasture remains firm no matter how much rain we get. We just need to manage the areas around the sheds and feeders.

There is something of a catch-22 here. Planning what kind of holding you need along with the land and buildings is all very well. But the chances are you will find something that doesn't exactly match your requirements. In other words, the holding you choose may determine what mix of animals you can keep. It is all a bit of a balancing act. It is, however, something to be aware of when choosing your smallholding.

One more thing to think about is safety. It is a sad fact that the largest threat to sheep is the domestic dog. And behind the domestic dog is an army of owners who believe their pooch is perfect. And they may be, until that moment when they see a sheep bolting and the chase instinct kicks in. All dogs have it. If something runs away, a dog cannot help but feel compelled to run after it.

So, when looking over prospective properties, bear in mind rights of way and access to the general public. Rights of way tend to be an English concept. In Scotland, the right to roam exists. In essence, that is fine, but there will always be a minority who think it is fine to wander through flocks or herds of animals not realising they are causing widespread panic. Or, in some instances, not even caring.

Access from roads is also something to think about. As I write this, there is a growing trend for criminals to butcher animals, especially sheep, in their fields and take the meat away. You are left with the horrifying find of a discarded fleece. In some ways, this is an argument for going more remote. These criminals will go for easy pickings first, so that means near main roads and built-up areas.

Last but not least, there is the subject of money. In

other words, can you make a living off a smallholding? The answer is that it is not easy. You can, maybe, grow enough food to sustain a family, but it requires careful planning and hard work. But there are still bills to pay. In the UK, there's council tax along with – depending how off-grid you plan to be – water, electricity, oil and heating bills. Add to that car running costs, insurance, house and general maintenance, and the costs are soon mounting.

Generating an income is possible, but hard. Many farms and smallholdings now run holiday homes. Many farmers have jobs alongside farming, for example tractor work and fencing. When we made hay, we hired another local farmer to come in and cut and then bale it for us.

As with many smallholders, we both have jobs separate from running the smallholding. Nicole is a gardener, and I teach guitar and work as a mathematics tutor.

Later chapters offer example costs and some ideas for generating income. These cover ideas such as crafts, breeding livestock, selling eggs and honey, and producing meat.

So, if your goals include self-sufficiency and/or generating an income from your land, you will need a plan. The last thing you want is to be stranded on a smallholding with no money and bills to pay. This may sound a bit scary, but we have found from our own experience, and on many of the TV programmes featuring people moving to the country, that a common theme is we all end up with a lot less money than in our city lives.

It can all sound a bit doom and gloom. It is not. The idea behind this book is to help you think ahead and plan your smallholder life. Get it right, or even close to right, and you will save yourself a lot of time and effort.

It is hard to get the balance right. In business life, strategies can be planned (think 1970s Soviet economy) or emergent. Emergent strategies are basically actions that are developed over time in the absence of any particular plan (kind of like making it up as you go along). In planning how to get your smallholding right, you will most likely need to combine both approaches. Just beware that not enough forethought could mean you find yourself putting in fencing and infrastructure only to find yourself ripping it out and doing it again. For example, you could find yourself falling in love with Highland cows, buying some, and then realising not only how expensive they are to keep, but also how much time you are spending repairing fences.

There are a lot of things to consider. We used a picture board to help focus us. On it we placed pictures of all the things that were important to us. It helped us a lot and was also quite fun to do.

For those of you who like to be a bit more organised, you could always set up a spreadsheet. This could look something like the one shown next:

Must Have	Importance
at least two bedrooms	10
Seven acres of pasture	10
drinking water for animals	10
dry areas in winter	10
vet within one hour	10
no public footpaths	10
central heating	9

Nice to Have	
woodland	7
Sheep-handling area	6
lambing shed	5
vegetable patch	4
outbuildings	7
away from main road	8
shop within ten minutes' drive	7
not too remote	7
close to but not in village/town	6
max two hours from regional airport	4
Aga / range cooker	6

Each property could be compared to the above spreadsheet and scored on these important factors. When you have added up the scores for each property, you can see more clearly which is best suited to your needs.

At the same time, you can ask yourself whether it feels right. But, if you do only this, it would nevertheless be a good idea to ensure it meets the "must-haves"; otherwise, you could come to regret your choice later.

5

LARRY AND LISA

We were still in Somerset, and it was our first time lambing. Nicole had spent hours poring over books and reading articles on the internet. "This is not going to be easy," she kept saying. In fact, what she meant is that it was going to be really tough. I have to admit I was a bit more laid back about it. Ignorance is bliss. Well, yes it was. But it didn't last.

We had nine sheep, and so far as we knew they were all pregnant. We'd borrowed a tup, Ginge, from our mentor, and he had proved a popular chap with the ladies.

During tupping, Ginge had worn a raddle containing a block of paint. Every day, we inspected the ewes and once they had a paint mark on their rump, we made a note of their name and date. Through this, we built up a diary of when the births could be expected. This is a really useful list. It is not entirely accurate, but it does give a good idea of what to expect. Sort of!

Lambing was fast approaching. It was early February and our first lambs were due in mid-March. We had our girls in the field closest to the house so we could keep an eye on them. It wasn't the biggest field, but being winter they were eating hay, so the amount of grass wasn't really an issue. They also had two small field shelters. Although, like most sheep, they like to stay outdoors unless it is really tipping it down.

That winter, the weather became relentlessly wet. The ground was pretty flat, so it soon became waterlogged. Sheep are pretty hardy animals, but even they were getting a bit fed up with the constant rain.

The problem was that they couldn't tell us, and we were too inexperienced with sheep to read the signs. Until the really big sign was posted.

One morning, we saw a small black mound in the field. We went over to investigate – it was a lamb. The poor wee thing was dead. It was devastating. Our sheep are really important to us. Their well-being comes before anything. And here was a wee lamb that never got a chance to live and blossom. Worse, we found another. And her poor mother (Peaches), after all that time raising them in her womb, was left bereft.

This was a wake-up call. Something was not right. Aside from this being devastating for us, spontaneous abortions are bad news in a flock of sheep. If one goes, it can start a chain reaction. We had to do something, and we had to do it quickly.

We consulted the vet who suggested bringing them in early as the abortion might have been triggered by stress because of the awful weather.

We had prepared a shed ready for lambing. We had originally built it to house the tractor and trailers, but it was the only decent shed we had, so the tractor and trailers had been parked outside. I had also laid a concrete floor – the books had said this was best. That's not necessarily true, but at that time it was the best advice we had.

It was rubber gloves on, and we scrubbed the shed down with disinfectant until it was spotless. We set up water and power and got ready to bring in the sheep. However, it was not that simple. In an ideal world, we would separate Peaches so she could not inadvertently trigger abortions in the others. But that would have meant both leaving her out in the appalling weather and also, worse, on her own. Sheep cannot cope with being on their own.

We had no choice; we had to bring Peaches in with the others. It was a worrying few days.

Ideally, you bring the sheep in at the last minute for lambing. You want the lambing environment to be clean, but it is a lot of work to keep clean when it's full of sheep all pooing for Britain. We had brought them in five weeks early. We did a lot of mucking out. That said, the sheep were ecstatic to be out of the rain. With plentiful fresh hay on offer and drinking water on tap, they settled down and watched the rain from the comfort of their own shed.

The weeks passed and the weather remained poor. As the first due date approached, we started the four-hourly checks. We had been advised that four-hourly checks were frequent enough to catch any problems (these days we do two-hourly checks). But even four-hourly checks every day and night soon wear you out. We were both working as well as lambing, so, day by day, we became more and more sleep deprived.

Eventually, something happened. One evening as the sun was setting (around six o'clock) Scarlett went into labour. At least, she was showing signs of going into labour. An hour later, she was still in labour. Was she struggling? We couldn't tell, so we called the vet. Although we were in rural Somerset, all the local vets only handled pets, so we had to wait for a vet to come up from Devon. That was a long wait. It is unbelievable how long an hour can feel when you have an animal in distress.

Eventually, he arrived and diagnosed a tight cervix. This was a little surprising as it was not Scarlett's first time. Nevertheless, that's what it was. The unborn lambs were basically trapped. The vet helped them to be born. A boy and a girl, who we called Larry and Lisa. They were so cute, those little black bundles of wool. So cute that we took our eyes of the ball. We did not get them in front of Scarlett quickly enough. With all the struggles she'd had plus the presence of the vet, Scarlett was in a bit of a state. She made a run for it. We brought her back and presented her with her lambs. She was having none of it.

Scarlett was so stressed she rejected her lambs. This was not going well. Two dead and two rejected. This was, as Nicole had predicted, turning out to be really hard. We did everything we could to bring Scarlett, Larry and Lisa together. We spent hours with them, helping them to suckle. But Scarlett kept butting them away. Just after midnight, Larry went over and lay down in a corner. It would have been easy to assume that he was just having a wee sleep. But there was something in his body language that rang alarm. It felt like he'd given up.

We made a decision — we would hand rear Larry and Lisa. To be honest, we were a little thrilled at this. It would be like having pet lambs. That's what it is like when you take on something new. It is exciting. You have no idea what you are letting yourself in for.

Larry and Lisa were a joy. We set up a feeding programme based on a mix of two-hourly feeds during the day and four hours at night. As we were already getting up every four hours, this was no real effort. And the contentment you feel sitting on a chair at three in the morning with a wee lamb guzzling milk on your lap is hard to describe. Getting up at three became something to look forward to.

We had a pen set up in the living room so they would stay warm. Once they were strong enough, we would put them in a pen in the lambing shed during the day so they could be surrounded by sheep. It was not uncommon for one of us to find the other fast asleep in that pen with the lambs curled up on our legs.

In the evenings, they'd skip around the house, exploring, before curling on our laps in front of the TV.

As they grew, and as other lambs were born and put out to pasture, we were able to set them up outside. It is important for lambs to have access to grass very young. It helps their rumen to develop. Once we had a few lambs outside. We set up a Larry and Lisa pen in the middle of the field. This had shade and shelter. By now we had them feeding from bottles on a rack, so they were able to feed themselves. We kept them from the other sheep at this stage, as we were not sure how the others would react to Larry and Lisa.

We continued the night-time feeds for weeks after all the lambs were born. It was hard to know when we could stop. It was also hard to know when to let them integrate fully with the others. But, somehow, we managed it. One sunny day, we opened up the pen and sat and watched. In no time, Larry and Lisa were amongst the other lambs and they were all fine. Lisa turned out to be quite the little playgirl. She was often seen skipping and jumping round the field. Once one starts, others join in. Before long, you have a flock of lambs all running and jumping. Watching lambs playing is one of the things that makes lambing worth all the effort.

Larry and Lisa were a success story. Larry went on to take two first prizes at the Bath and West show. He lives on a small farm in Somerset and produced 16 lambs in his first year.

Lisa also moved with two of her half-sisters to live on a smallholding in Somerset where she leads a pampered life.

6

PLANNING AND BALANCE

To some, the idea of planning can bring a warm glow. To others, the very thought is enough to bring them out in a cold sweat. Fear not, this is not a chapter about how to build a detailed plan for your smallholding. As it happens, I was a project manager in my corporate life, so I understand the intricate nature of project planning. However, that sort of plan does not really work for smallholdings. If anything, it can create more stress and problems than it resolves. What is really needed is the answer to two questions:

What do you need?
When do you need it by?

The "what" will largely be dictated by budget, space and how self-sufficient you wish to be. The "when" will be determined by time, the natural cycle and, to a certain extent, budget.

The assumption here is that you have found your smallholding and moved in. If you have not yet moved in, you are not far off, or have access to look around in advance of moving in.

It is also worth bearing in mind that things change. When we chose our first smallholding, it was our goal to keep chickens and grow vegetables. Keeping livestock never featured in our thoughts. Yet within a year we had sheep, and within two years we had run out of space.

But it is a good idea to draw up a list of what you want to do.

This list will likely be a combination of things such as vegetables, fruit, chickens, bees, sheep, pigs, and so on. You may also be looking at growing wood if you have woodburning stoves. For each of these, you will need to work out how much space you need.

There are individual chapters for each of the above and more later in this book. In these chapters, you can find details on how to work out what space you need. For example, it will help you to work out the number of sheep per acre, and so on.

A key thing to think about as you build your list is time. This is not about when you need it by. It is more about how much time you have. The more you want to do, the more time you will need. Smallholding is a tug of war between wanting to be as self-sufficient as possible, having the time to look after everything and the need to generate money.

We talk about money in more detail in another chapter. But the fact is, you cannot live for free in the UK. At the absolute best, you might be able to strip out all costs relating to food and heating, but there are always bills to pay such as animal medication, insurance, council tax and transport.

The thing is, to grow enough vegetables to feed yourselves and to keep animals along with general maintenance is pretty much a full-time job. It is hard to generate income from a smallholding. All the things that you can produce are low-value items. You can make money, but you can't make a living. Everyone we know living this life has a job outside of the smallholding in one form or another. And jobs take time.

So, part of the planning process is to work out how much money you need in order to survive and how much time it will take you to earn it. This will give a rough idea of the time you will have left to run the smallholding.

I would say that at this stage you don't need to go into too much detail. Also, the chances are that you have some experience in some of the things you are going to do. For example, I had run an allotment for a few years so had experience in growing vegetables. I had also kept chickens in my back garden. Nicole has an RHS qualification, so we were well prepared for all things horticultural. Or, so we thought.

So, having worked out your outline list of goals, the next stage is to plan the layout of your smallholding. A lot of this will depend on what's already there. If you are lucky, there will be a set of well-maintained outbuildings along with well-fenced paddocks. What's more likely is that there will be one or more paddocks fenced off with ageing fences and a few outbuildings in various states of disrepair.

It is worth thinking about this carefully. By this, I mean what goes where.

For vegetables, you need to have a large enough area that will enable you to plant everything you want to grow and still have space left to leave areas fallow. There are plenty of excellent books on how best to grow vegetables. It is not the intention of this book to cover that. Nevertheless, there are a few key things to think about when planning a vegetable garden.

By growing vegetables you are, in effect, creating an area of monoculture, and with that comes a range of challenges. Most of these can be grouped under the title "Pests". And believe me, pests can devastate an entire planting. Pest management benefits from having an area left with no produce growing on it each year. This can simply be mulched or you can grow nitrogen fixing "green manure" such as clover.

If you are aiming for self-sufficiency, you will need quite a large area set aside for growing vegetables. You will also need to think about seasons insofar as you will want to have the vegetable garden productive all year round. It requires careful planning to have vegetables available in the early part of the year. This is easier in the south than the north.

In the north, all-year round productivity is likely to need a polytunnel. Wherever you are, growing from seeds benefits from having a greenhouse or polytunnel.

The final point on vegetables is choosing the best site. Your vegetable garden will ideally be sheltered yet have access to sunlight all day. Damp, shady areas do not work well.

If you are planning to keep animals, each has a specific set of requirements. These are covered in later chapters. In terms of planning, there are some general aspects to consider.

The first is to look at how much space the animals need. There is plenty of advice on the recommended numbers of animals per acre on the internet. What I would advise, whatever you do, is don't overstock. The actual number of animals your land can support will depend on the quality of the grass and the type of land. Even a few animals can turn a grassy field into a mud bath if you have heavy rain.

It will also depend on whether you are planning on breeding your animals. If you are planning to breed, you will need to ensure you have enough space for the young animals, too. It can be all too easy to stock your plot, breed some lambs and then run out of grass. This is particularly true of cows. If, for example, you have two cows that you're breeding for milk or meat, you will need space for four or even six cows (the mothers and their calves). That's a minimum of an acre each, plus, if you are cutting your own hay, an acre each for hay. That's 10 to 12 acres already allocated.

Some animals require winter shelter. Most breeds of sheep are pretty hardy, but most breeds of cows need to come in for the winter. And while sheep are hardy, in our experience a field shelter is a must-have. It is also generally considered good practice to move animals around. Too many animals in a small space can lead to disease, one of the main problems being worms.

If you are keeping males and females, this will need to be managed. As a rule, the boys and girls are kept apart except at breeding times. If you do breed, the boys need to be separated at a certain age. All this takes space.

You will also need to inspect the fencing. A motivated sheep, tup or ewe can breach a stock fence if it really wants to. You only need to look at a cow or a bull to find yourself asking how a stock fence ever keeps them in.

We have found that chickens really thrive when given enough space. Chickens can be pretty cruel to one another. Anyone who has introduced new chickens to an existing flock will know how aggressive they can be.

The more space they have, the easier it is for bullied hens to get away. So, the more space they have, the less of that pecking and bullying behaviour goes on. Ideally, you will be able to let them range freely. This will depend on the proximity of other properties and roads. However, like most animals, once chickens have established their territory, they tend to keep within it. Ours are fenced in, in quite a large area. But there is an open gate into the fields at one end. They rarely venture out, as their main area is more than enough space for them.

In our opinion, the generally accepted ratio of one chicken to one square metre is nowhere near enough space.

We have also found that pigs need much more space than what is often quoted. We built what we thought was a generous run. Within a few weeks, we could tell they were bored, and they had turned the area into a quagmire. Fortunately, we had the space to build an extension, so we doubled the size of their area. When we let them into the extension, they loved it. Pigs also need shade. As well as being woodland animals and, therefore, naturally seeking to root around under trees, they are easily sunburnt. Trees give the best shade. In addition, pigs need a wallow, so all of this combined needs a fair amount of space.

With animals, there is also a set of equipment you need. Animals get sick, therefore you will need management areas where you can control them in order to administer injections, and so on. These can range from a simple pen for sheep through to a pig board and corner for pigs. Cows need specialist equipment. There are many devices out there purporting to help with managing animals. Don't rush and buy these. Aside from being expensive, most are targeted at commercial flocks where time rather than animal comfort is of the essence.

We have found it a much better strategy to earn the animals' trust, and so minimise the stress, effort and equipment needed in order to administer medication. You can read more about this in the animal chapters.

Having made your list of what you plan to do, you should find that there are a number of things to be carried out under each subject. It is often a long list and can appear quite daunting.

Our approach is to group the jobs into two categories: "urgent" and "not urgent". Some of this will depend on the time of year. If you have moved into your smallholding in January and need to build a vegetable garden, then you will need to get cracking. Planting out can begin as early as April.

If you are planning to buy animals, you can, as a rule, buy them at any time of the year. However, the normal time is during the summer months. You will also need to ensure the animals have access to fresh water.

If you are planning to breed animals, you will need to have the above-mentioned facilities in place before the first lambs or calves are due.

Considering the natural cycle, i.e. when animals are born and when to plant vegetables, fruit bushes and trees, etc., will help give a rough idea of what you need to do and when.

Putting all of the above together will help you build your general plan.

If it is anything like ours, there will be a lot of things to do. This is when you will need to think about how much time you have and also what your budget is.

If you are pushed for time, you can always subcontract the work. In rural areas, there are usually local contractors who can do fencing and structural repairs, and so on.

It is worth thinking about this. It can be very easy to throw money at the problems. Sorting out fences, buying animals and buying equipment can be expensive. A little forethought can go a long way and save a bit of money, too.

One final thing to think about regarding money. All these things cost money to run. With vegetables, there are seeds to buy and fertilisers. Animals require winter feed. Chickens need to be fed all year round. If you keep pigs, unless they have acres of woodland to forage in, they will need to be fed daily. It is also highly probable that you will need to call out the vet from time to time. So there are many ongoing costs to consider.

Assuming you are planning to generate revenue, you could be thinking about selling eggs, honey, or even breeding and selling livestock.

Throwing all of the above ideas into the pot should give you a good general idea of what you need to do and when.

In our experience, it can take two years to sort out a smallholding. In our current place we built a pig area, a field shelter, a water capture system comprising four water tanks and pipes to troughs in every field, a lambing shed with an electricity supply, a lambing paddock, a sheep-handling area, a cow-handling area, a hay shed, and also sorted out a number of fences and drystone walls. We also constructed a large vegetable patch alongside the existing one (which had seemed quite large when viewing the place but turned out to be quite small when it came to planting).

In our experience, it is also very easy to overestimate what you can earn and underestimate what it will cost.

So, part of the planning process is to make adjustments as you go. As you gain experience of both your patch and your animals, you will probably find better ways of doing things.

Our goal is to make the best possible area for our animals while at the same time making things as simple as possible for ourselves.

The true art of the smallholding life is finding the right balance between self-sufficiency, time and generating enough money to survive. You can't really plan for this. You can set yourself goals, but in the end experience will dictate the best balance for you.

Next are some sample ideas on calculating what you can keep on your holding (in relation to pasture size). Note: time and expenses are covered in the individual chapters for each animal.

The following table shows calculations for a smallholding with seven acres where you would like to grow vegetables, breed sheep and keep chickens.

						acres
Vegetable garden						0.5
Animals	per acre	no.	breed	off-spring per animal	total animals	
Sheep	3	7	yes	1.6	18	6.0
Cows	0.5	0		1	0	0.0
Pigs	24	0		9	0	0.0
Chickens	80	10		1	10	0.1
Total Acres						**6.6**

The above shows that taking breeding into account along with an average of 1.6 lambs per ewe, you could keep a maximum of 7 sheep.

A further example shows a smallholding with 13 acres and you would like to grow vegetables, breed sheep, and keep chickens, cows and pigs.

	per acre	no.	breed	offspring per animal	total animals	acres
Vegetable garden						0.5
Animals						
Sheep	3	9	yes	1.6	23	7.8
Cows	0.5	2		1	2	4.0
Pigs	24	2		9	2	0.1
Chickens	80	20		1	20	0.3
Total Acres						**12.7**

The above shows that taking breeding into account along with an average of 1.6 lambs per ewe, you could keep, for example, a combination of 9 sheep. 2 cows, 2 pigs and 20 chickens.

Note: the above numbers of animals per acre are based on the UK's requirements for organic farming along with personal experience. It also depends on what happens to the offspring – for example, if they are all to be sold in the first year, then stocking densities could be higher.

7

YMOGEN

One Saturday morning, Nicole noticed that Ymogen, one of our lambs from the previous year, looked slightly out of sorts. She was standing slightly away from the others, head drooping. Nicole went to check and found that Ymogen seemed to have something very wrong with her mouth. The bottom part was all wobbly and seemed to be detached somehow; it was kind of just flapping.

We called out the vet, but deep down we both had the same feeling: Ymogen could not survive such an injury. If she couldn't use her mouth, how could she eat? Would she require surgery? And so on. Thankfully, there was a vet available, and she set off straight away. Nevertheless, it was another of those waits where half an hour seemed to stretch on forever.

The vet confirmed our worst fears. Ymogen had broken her jaw in such a way that there was nothing she could do. For a while we stood there doing nothing, resigning ourselves to the worst. We really didn't want to ask the vet to put Ymogen down. While the vet was waiting for us to decide what to do, she suddenly offered up, "She's young, you never know, she might survive."

Well, that was all the encouragement we needed. Ymogen was given a painkiller and some antibiotics. Following on, Ymogen would need an injection of antibiotics and painkillers every two days until she was out of the woods. The vet supplied us with bottles of the relevant drugs, and we drafted a treatment plan.

However, we still had to keep Ymogen alive, somehow. We retired to ponder what to do. It is strange; under such circumstances, your brain seems to stop functioning correctly. Perhaps emotions trump logic. Who knows? We were trying to work out how to get Ymogen eating again but getting nowhere. Inspiration came via one of our neighbours. She, on hearing the story, offered some of her guinea pig's chopped premium hay. Chopped hay! What a brilliant idea. Why didn't we think of that? It is not like we hadn't used chopped food in the past.

As well as chopped hay, we looked into high-energy foods for sheep. We settled on carrots, turnips (swedes outside of Scotland) and parsnips with a sprinkling of Brussel sprouts. Luckily for us, we had plenty of these in our vegetable patch. It was February, so we were indeed fortunate to have had a bumper year for winter vegetables.

Nicole prepared a masterpiece of a meal for an injured lamb. Shredded vegetables. We took it in turns to head out every hour or so to see if we could get Ymogen to eat some of this. We made it a little more tempting by sprinkling some powdered sheep nuts on it. We had to be careful here. With Ymogen not eating grass, we needed to make sure she did not eat too many sheep nuts. There was a high risk she might eat the nuts and nothing else. An overdose on these high-protein nuts can lead to acidosis, which can also be fatal.

Sheep don't make it easy!

At first, Ymogen was reluctant to eat anything. She was probably still in shock. It must have been quite a painful injury. We had spent a lot of time trying to work out how it had happened. The day before, we had been doing the annual vaccinations of Heptavac. To do this, we bring the sheep in one at a time into a pen, and I hold the sheep while Nicole administers the injection.

We're not sure, and it is likely we'll never know for sure, but it is possible that in all the comings and goings Ymogen, still being small, had put her head through the bars and then been hit from behind by another sheep. It makes my toes curl just thinking about it. We have, since then, designed and built a permanent sheep-handling area using stock fence instead of hurdles so this can't happen again.

Anyway, Ymogen did lick up some food on those first feeds. Maybe two or three handfuls. So, a strategy of little and often emerged.

One of the side effects of sprinkling nuts on Ymogen's feed was that it attracted other sheep. This meant we had to try and get Ymogen *into a pen to feed her. This was easier said than done because, with all the injections she needed, she was beginning to associate the pen with painful pinpricks. We persevered and, over time, it got a bit easier.*

After a couple of days, Ymogen had still not really eaten very much. It was now Monday and Nicole had gone to work. I work from home, so I was able to keep up the regular feeds. I had prepared the usual amount: about a quarter of a small bucket full.

As usual, I was having trouble tempting Ymogen into the pen, but, for once, the other sheep, Yogi aside, were showing little interest. Yogi is about the same age as Ymogen, so that wasn't too much of a problem. It is the large sheep bumbling in that you want to avoid. So, with Ymogen half in, half out of the pen, I offered her some food in my hand. She ate it all. I offered her more. She ate it all. In fact, she ate everything I had brought. It was an amazing moment, hard to describe. I texted Nicole. Her response "huh?" summed it up completely.

I went and got more food, and she ate half of that, too. I think that, for the first time, we really thought there was a chance she might get through this. After that, Ymogen did start to eat more. Our optimism grew.

Given the nature of her injury, Ymogen could only eat slowly. Each feed was taking about 45 minutes, as she had to lick it all up from our hands. Being February, we were feeding Ymogen in snowstorms, hailstorms, howling winds and generally inclement winter weather. We often found ourselves standing over her protecting her from the elements while she ate. The sheep do have a winter shelter, but as a rule they prefer to be outside.

Some feeds had to be done on the move. Ymogen would panic if the other sheep moved off without her. It would be a handful here, then lead her to the flock's new location, then another handful. Then another walk. And so on. She was on five or six feeds a day, so we were spending a lot of time outdoors in the winter weather. We got some pretty cold hands, let me tell you. But it was worth every second.

Over the next two weeks, Ymogen became able to eat more quickly. She also started to eat out of the bucket by herself. That made things a lot easier for us insofar as we could keep our hands warm.

We were also able to leave Ymogen in her pen to eat while we did the general sheep checks. What had become a full-time job slowly became more part-time.

Ymogen had also learnt that she had nice meals when she came into the pen. She started waiting for us and skipped into the pen quite happily. In fact, Ymogen has always been a happy sheep. She just has one of those naturally happy faces along with a generally happy demeanour. Watching her recover is one of the most rewarding things we have ever experienced.

After around four weeks, we were able to stop the injections. To be honest, Ymogen seemed to have got used to them. With her head in a bucket of food, she hardly flinched when the needles went in. She's a brave wee soul. But it was good to see her eating without painkillers. She was definitely on the mend.

After about six weeks, Nicole saw Ymogen grazing. Ymogen was actually eating grass. She was so happy she filmed it. We continued to offer Ymogen her turnip and carrot mix, but over time she went back to eating grass. Her jaw has healed. It is not perfectly straight, but it is straight enough that she can function as a sheep.

Having had all that contact with us, Ymogen always runs over to us now when we go out to check on the flock. All our sheep are fairly tame and tolerant of us and our dogs. But a few of them actively seek out our company for pats and scratches and general attention. Ymogen is at the top of that little group.

We still get the impression that the vet can't believe Ymogen survived. While this is kind of satisfying in itself, nothing can beat the warm glow we feel each time we see Ymogen.

Ymogen is a miracle sheep.

8

SMALLHOLDING AND MONEY

A hundred years ago, small farms were the norm. From crofts across Scotland through to small farms in England, Northern Ireland and Wales, people lived and worked the land. They worked together sharing resources and produce.

Look now and you will see that most of these crofts and farms have gone. Some are now under housing estates, some have been absorbed into larger farms and some lie derelict.

For some reason, in the last hundred years or so this way of life became unsustainable. The reasons behind this are complex, but one of the main factors is the rise of the supermarket. Supermarkets have pushed food prices down and at the same time changed the way food is presented. As a child, I remember all vegetables came covered in dirt as if they had just been dug from the ground. We had to wash everything. We had implements to remove the bad bits.

These days, everything is washed and presented as near perfect, often wrapped in plastic.

Supermarkets are supplied by large agricultural firms, mainly large "factory" farms. There are few places left for small farmers to sell their produce.

Increasing levels of mechanisation and automation means that large farms can now be run by a small number of people. In comparison, small farms and smallholdings are labour intensive. In economic terms, this means the costs are higher.

This chapter looks at the reality of smallholding life from a financial perspective. First, the bad news, there are always bills to pay. Below is a table showing a representative sample of costs and an annual budget. Please note that all the figures used in this chapter are based on 2020 prices.

One column shows an "average" household with all water and energy sourced from utility companies and some food grown. The "high self-sufficiency" column shows what may be achieved if you are on a private water system, a septic tank, all electricity supplied by solar panels or equivalent, and all heating and cooking from home-grown wood. The latter also assumes high proficiency in mechanics insofar as you can service all of your own equipment including cars.

Expense	Average family	High self-sufficiency	Assumptions and sources
Council Tax	£1,925	£1,500	*four-bedroom house connected to mains water and sewage*
Food	£4,750	£600	*source Google*
Phone / internet	£360	£360	
TV Licence	£155		
Electricity	£1,250		*source Google*
Heating	£1,250		
Water	£400	£200	*private supplies need new filters and ultraviolet lamps each year*
Farm Insurance	£500	£500	*this can vary depending on what you insure*
Service farm equipment	£450		

Expense	Average family	High self-sufficiency	Assumptions and sources
Animal feed	£750	£750	*assuming around 20 sheep and 10 hens*
Dog food	£300	£300	*assume one medium-sized dog*
Vet	£360	£360	*based on our experience*
Clothes	£500	£500	*things wear out, especially jeans and boots*
Service boiler	£150	£150	
Car maintenance	£750		*this will depend on the age of your car*
General maintenance	£500		*a rough guess*
Quad/tractor fuel	£200	£200	*depends on how much you use it, but quad bikes get used a lot*
Car insurance	£200	£200	*assume one car*
Car tax	£300	£300	*assume one car*
Car fuel	£1,371	£1,371	*assume 8,000 miles at 35mpg*
Total annual cost	**£16,421**	**£7,291**	
Per month	£1,368	£608	

As you can see, even if you eliminate all fuel costs and limit food shopping to £50 per month (for stuff you must have but can't grow), you could still be left facing bills of around £600 per month. It may also be possible to reduce your costs further by, for instance, living in a hut or caravan rather than a house. This would reduce insurance costs, for example.

Bear in mind that the above is purely illustrative; the idea is that you would create your own table of costs depending on your specific situation.

Furthermore, the figures in the previous table does not include mortgage or rent costs. It is hard to apply a figure for these as each case would be individual. Also, prices on land seem to vary quite considerably across different locations. Needless to say, if applicable, you should include mortgage or rent costs in your own calculations.

The previous table also does not include start-up costs or replacement costs. These are also things to think about. Start-up costs are detailed in later chapters for the different types of animals.

If you are looking to give up your jobs and earn your living from the smallholding, then here are some ideas of what you can sell and for how much. Actual numbers would be based on the size of your holding and the number of animals you can keep, produce you can generate and other possibilities.

Item	Cost[1]	Sale Price	Profit	
Rear and sell lambs	£160	£100	-£60	factors in costs of extra feed for ewes, likely vet costs and vaccinations, and annual cost of keeping breeding flock
Rear and sell calves	£320	£400	£180	
Rear and sell piglets	£30	£40	£10	
Rear and sell lamb as meat	£165	£150	-£15	
Sell wool fleeces	£2	£1	-£1	
Sell honey from one beehive	£120	£360	£240	assume average year producing 18kg and selling for £4.50 a jar
Sell eggs	£24	£60	£36	assume £2 per dozen, one chicken will lay 250 eggs per year
Sell wool as yarn	£32.50	£83	£51	assume six balls of yarn per sheep, costs include shearing and spinning, sale price £9 per roll

[1] *Cost means the cost of production, for example the cost of producing a lamb includes the cost of keeping the ewes, hiring a tup, extra feed, medical costs, vaccinations, and so on. More details on costs can be found in the chapter on each of the above later in the book.*

There are many other ways to earn money. The previous table is based on a sample to show that the basic items produced on smallholdings do not provide a lot of income.

Ways to generate more income are generally based on adding value. I'll talk about those later. For now, looking at the previous figures, we can make some forecasts.

The following table assumes the high self-sufficiency route with no rent or mortgage to pay. It calculates the numbers of animals you would need in order to make the target income of £7,291 per year (as defined in the previous table).

Rear and sell calves	48	healthy calves each year	96	acres needed
Rear and sell piglets	720	healthy piglets each year	5	acres needed
Sell honey from one beehive	38	hives needed		
Sell eggs	114	chickens needed		
Sell wool as yarn	142	sheep needed	29.5	acres needed

Note: table excludes sheep, as based on the above figures you cannot make money from sheep.

From the above table, the breeding of pigs would seem the only route to break even on the average smallholding. And that is fraught with difficulty as it might prove hard to find buyers for 720 piglets.

Looking at the average household with expenses of £16,421 per year (from the table above), we get the following table:

Rear and sell calves	109	healthy calves each year	218.7	acres needed
Rear and sell piglets	1640	healthy piglets each year	11.4	acres needed
Sell honey from one beehive	86	hives needed		
Sell eggs	259	chickens needed		
Sell wool as yarn	323	sheep needed	67.2	acres needed

Note: table excludes sheep, as based on the above figures it is nigh on impossible to make a profit from breeding sheep and selling lambs.

Adding rent or mortgage payments of £1,000 per month gives the following table:

Rear and sell calves	189	healthy calves each year	378.7	acres needed
Rear and sell piglets	2840	healthy piglets each year	19.7	acres needed
Sell honey from one beehive	149	hives needed		
Sell eggs	448	chickens needed		
Sell wool as yarn	559	sheep needed	116.4	acres needed

Note: table excludes sheep as based on the above figures, it is nigh on impossible to make a profit from breeding sheep and selling lambs.

The purpose behind the above tables is not to give a definitive answer on how to make money from a smallholding. For a start, livestock prices fluctuate on a daily basis. Different breeds have different costs and prices. The tables are only intended to show that is very hard to make enough money even just to pay the bills. This kind of fits in with demographic changes over the last hundred years. Small-scale farming has been largely replaced by large industrial farming. To make a living, you need to keep a lot of animals and be ruthless about costs.

You can increase your sale prices by focusing on rare breeds and producing prize-winning animals. A prize bull or ram can fetch quite a lot of money. However, prices are market driven and can fluctuate wildly. It can be a bit of a gamble.

One of the things we thought about was that if we did choose to go down the route of generating all our income from farming, we would, in effect, be returning to a form of corporate life with its associated balance sheets, stresses and business dealings. We did not want to have shops breathing down our necks for more stock.

A further sobering thought is that it is actually quite hard to sell high-quality produce from smallholdings. We have heard stories of smallholders selling their produce in a local shop but finding that hardly any had been sold when they went to restock.

The fact is, most people whether in the country, towns or cities tend to shop in supermarkets. There are farm shops around, but it can be hard to get your produce into these.

You may have noticed that I have not mentioned selling vegetables. This is a possibility. However, you have to bear in mind that people are now used to buying sanitised vegetables. Even people buying produce from farm shops don't want to find slugs and caterpillars lurking in their vegetables. So, if you want to sell vegetables, you will need to wash and prepare them. Then there's the task of finding somewhere to sell them. Remember, you are needed at the smallholding; you can't be sitting in a shop all day.

This may seem a pretty bleak picture. The fact is, it is hard to make a living at farming without scaling up.

One answer is to find a job. As I have already mentioned, everyone we know leading this life has some sort of job. Whether it is working in the local shop or bar, there is usually something you can find. It is better if you can use your skills to set up a business. Around here, tradespeople are always busy. It can take weeks to get an electrician, plumber or carpenter. Smallholders and country dwellers are always on the lookout for good mechanics, people good at fencing, roofers, builders, tree surgeons and general handy types. If there is a town nearby, then you can make quite a good earning.

Smallholders are also always on the lookout for good sheep shearers willing to shear small flocks. We have also, in the past, hired local farmers to cut and bale hay for us.

Myself, I teach guitar and travel into a large town nearby where I hire a room in which I can give lessons.

So, when planning for your smallholder life, it can be a good idea to factor in some kind of job. It is even better if you can factor in a job you can fit around your smallholding duties.

A lot of smallholders and farms also set up holiday homes. With Airbnb, this can be quite a good earner. This may not be possible if you are renting. If you own your smallholding and have the space, it can be an option. It might be easier to convert an existing building than build a new one. However, this is something you can check with your local planning department. Either way, it is something to consider and also something to cost up carefully.

The final thing to consider is a means of adding value. Take sheep for example, you cannot give the fleeces away. Converting them to yarn is expensive (unless you have the time to spin the wool yourself). And then you find your wool costs more to make per ball than what you buy over the internet. So you are selling an expensive product. It is not easy.

By adding value, what I mean is converting your raw produce into something valuable. You may be able to sell a ball of yarn for a profit of £4 or £5. But convert it into something like a hat, baby socks, a scarf, any knitted item, then you can sell it for much more. For example, if you make two hats from the above-mentioned ball of yarn and sell these for £12 each, and assuming six balls (twelve hats) from one sheep, your profit per sheep goes up from £50 to £112.

By adding value, you have doubled your profit which means, at the same time, you can halve the number of sheep you need. One thought, you might want to make more than just hats otherwise it could get a bit boring.

With the internet, these kinds of things can be sold online through any number of merchandise websites.

As well as knitted products, I have seen some smallholdings producing, for example, craft beers, craft ciders, craft jams and chutneys, and even cheese. Others are more specialised. One not far from here produces sculptures from scrap metal. These are much higher value items, so you may, with careful thinking, be able to generate more income from less. And that, really, is the key to making your smallholding pay. Find a high-value item that you can make and sell it.

9

ANT AND DEC

We had just moved into our new place in south-west Scotland and were still sorting out boxes. We had bought three ewes in lamb so we would not miss out on lambing. We were up at nights checking on them. To say we were busy would be an understatement.

Let's get pigs, we thought. As if we didn't already have enough to do. The problem was, I had visited a farm that kept pigs to see how it was done and it just looked so easy.

We had a pig arc and an electric fence, so we were all set. We found some weaners from a local supplier and ordered two. To be clear, these pigs were being reared for meat. It is not something we had done before. However, as we did eat meat and also had an ethical approach to food, we had decided only to eat meat from sources with excellent provenance. That meant local farms where we knew the animals were reared organically and on natural food (grass, not grain). It seemed the most ethical approach would be to rear our own. Then we could absolutely guarantee they'd had a good life.

We prepared the pig arc and set up the electric fence just as I had been shown. Then we set off to collect our weaners. This is where we made our first mistake. We have often transported lambs in the back of the car. Surely that would be fine for weaners. No! We should have taken a dog crate or something similar. The weaners were all over the place, climbing over everything and getting everywhere. In the end, Nicole had to sit in the back with them and keep them out of trouble.

So, we got them back and introduced them to their new home. It was a pleasant day, and their area was covered in fresh grass and flowers. The pig arc was full of fresh straw. They'll be delighted, we thought. We set them down next to the arc entrance and stepped back to watch. We named them Ant and Dec as we couldn't tell them apart.

They went off exploring. We saw them touch the electric fence and then move quickly back. It all looked to be working fine. After a while and given that we had rather a lot to do, we left them to it. We checked in every half hour or so, and it all looked good. And then, it went quiet. I nipped over to look. It was eerily quiet. I searched the run, but they had gone! I looked up and just caught sight of a tail disappearing through a stock fence into next door's garden. I yelled for help.

Nicole arrived quickly and caught the first escapee. Nicole handed her to me and set off after number two. I wasn't entirely sure what to do with escapee number one. I knew if I just returned her to the run, she'd likely escape again. In the end, I found some wood, put her in the pig arc and blocked the entrance.

By now Nicole and weaner number two (Ant or Dec) had vanished. I didn't even know which direction to go so that I could help. As I started wandering around looking for them, Nicole arrived looking totally dishevelled with a squealing weaner clutched to her chest. She had been over barbed wire fences, through brambles and bushes and finally, luckily, cornered Ant or Dec in another field and caught her.

We put her in the pig arc and had a conference. We had another pig arc door, so I fastened that on so we could keep them in overnight. We then decided to build a small run around the pig arc that was weaner proof. Together, we constructed a fence of chicken wire and put a strip of barbed wire along the bottom to stop them lifting it up.

Thankfully, that worked. However, it would only last a few days as we knew they would grow out of it quickly.

After researching pigs and electric fences, we discovered that pigs need to be trained to respect an electric fence. In essence, you need to put a barrier around the outside of the electric fence so they cannot go through it. Their natural reaction on touching the fence and getting a shock is to run forward. The barrier stops them going forward so that *they learn to go backwards. Hindsight is a wonderful thing; though, I did wonder why the pig farmers hadn't told me this.*

Anyway, we decided to build a proper pen using stock fencing. With pigs, the risk is that they will lift it up and go under, so it needs barbed wire around the base. Two strands is best. I set to work.

The first post went in easily. The rest did not. There were just so many stones at about 30cm down that it took days to knock in the posts. As I neared the end, with four or five posts left, I got careless and dropped the post stomper on my head. I was not knocked out, but I was bleeding a lot, so I went and got some ice. Nicole rang NHS Direct. I thought I was fine, but she insisted. The problem was, having just moved in and having no mobile phone signal, we only had a VoIP phone over a satellite link. This caused a delay of over a second. It was really hard to have a conversation. Needless to say, we now have a proper landline.

The net result was that Nicole drove me to hospital where they superglued my head back together and sent me home.

Next day, I was out finishing the fence wearing a hard hat. In between all of this, we were having problems with lambing (see the chapter on Vera, Vi and Violet), so neither of us were getting much sleep.

Nevertheless, we finally got the pen finished and readied ourselves for the opening ceremony. Just one little thing to do first...we had to tag Ant and Dec. Tagging animals is not something we like doing, but it is a legal requirement.

At that age, if you pick a piglet up, they act as though you are inflicting torture on them and scream for all they are worth. They are also already quite strong and very wriggly. Nicole did her best to keep them still while I put the tags on. It was, we felt, a "must get it right first time" activity.

Having already tagged sheep, we at least had some experience. Despite the tortured screeching and professional wrestler-level wriggling, we managed to tag them quickly. Only one tag each, thankfully. We also, in line with government regulations, tattooed them. That was pretty easy. You just tap them with a wire brush covered in ink (called a slapper). You buy the slapper with your flock number.

That done, we started to peel back the temporary fence. It was not long before Ant and Dec were out exploring. In fact, they loved all the extra space and started hurtling around the pig arc in big circles. It was great to watch. We felt, now, that we could relax a bit. Well, relax on the pig front anyway.

In fact, things did settle down; and Ant and Dec grew at quite a phenomenal rate. This also presented us with a dilemma. They ate a lot but did not have a large enough area to survive on foraging alone. We had to feed them, but the only food we could find was grain-based commercial pig food. What we really wanted was trailers full of vegetables. We never found a supplier for said vegetables, so we had to revert to the grain. Ant and Dec, however, did get a lot of fresh fruit and plenty of apples from our apple trees. However, their favourite was bananas, they just loved them, skin and all.

Over time, their fear of us gave way to friendship; they were soon presenting themselves for head scratches and the like. It is our philosophy that you have to get close to the animals so they can trust you. This means you can help them when things go wrong. While this worked well with the sheep and later the cows, it had mixed results with Ant and Dec.

One day, Dec (we could tell them apart now) started to look unwell. By that, I mean her behaviour changed, she became very lethargic and went off her food (always a bad sign). We were not sure what the problem was, but it looked like some kind of bug. While pigs are can catch bugs just like humans, we were puzzled. The weather had been warm and dry, and Nicole had kept their house clean with lots of fresh straw. Also, pigs don't foul their beds, so their pig arc was always pretty clean. When the vet came, we brought Dec into a small pen hastily assembled from sheep hurdles. The vet diagnosed pneumonia and gave her an antibiotic injection. He then gave us four syringes and said she'd need a five-day course.

No problem, we thought, that looked pretty easy. On day two, we brought Dec into the pen and gave her her injection. So far so good. On day three, we brought Dec into the pen. She was improving now and had a lot more energy. She took one look at the needle, sent the sheep hurdles flying in all directions and legged it. That was that!

Fortunately, two injections were enough and she recovered.

Over time, Ant and Dec turned their pen into something of a quagmire. They also started to look a little bored. We decided to build an extension.

I got to work. Armoured with a hard hat, I built a fence that would roughly double the size of their run. I put in a little opening so we could, if needed, keep the two areas separate. It only took a couple of days (I was getting better at this fencing lark).

In no time it was ready, and we got prepared for the second opening ceremony. The pig run area had a legacy concrete path, a sort of circle, running through it. I had positioned the gate so Ant and Dec could step out of the pig arc, trot along the concrete path and into the new run. We opened the entrance and called them. They duly popped out of the arc and along the concrete path into the new area.

They absolutely loved it. They started making happy pig noises as they found berry bushes, hogweed and also a handy tree to lie under. Of course, pigs are naturally forest creatures and their main run was pretty open. Another lesson learnt.

By now we had formed a strong bond with Ant and Dec. It was very hard for us when their time came. It is very easy to book them in at the abattoir. It is very easy to train them to go into a trailer. It is even quite easy to drive them there. But, it is ever so hard unloading them at the other end. The people at the abattoir were lovely. It is a family-run business, so that helped. But the drive back is when it starts to hit you.

And this is one of the problems we face. We want to eat ethically. We want our animals to live well. We avoid supermarket meat. We don't eat meat in pubs or restaurants. Ideally, we should be OK eating our own as we know the animals have been well cared for. But this experience has taught us that it can be terribly difficult. People say things like "don't give them a name". But, as already stated, our philosophy is to work closely with our animals so that they trust us. We find this makes it so much easier to handle them than relying on mechanical equipment such as races and crushes.

We went through with it. I made a ton of sausages and we had a freezer full of meat. But we also decided not to keep pigs again.

10

GETTING STARTED - INFRASTRUCTURE

Once you have moved into your smallholding, you can take stock of what infrastructure and equipment you have. Comparing this to what you plan to do should bring up a list of things you need. You can find information on what sort of things are needed for each type of animal in the specific chapters for each animal later in the book. This chapter looks at general things to think about regarding infrastructure.

An important starting point is to consider the layout of the land. What you are looking for is to identify where it is damp, dry, sunny or shady. Different things need different environments. You don't want your vegetable patch to be in a shady bog garden. Animals need both sun and shade. Animals also need areas where they can get out of muddy conditions.

Another thing to consider is rotation. Both crops and animals need to be rotated. So, your vegetable patch needs to have space to move things around. Your animals can also benefit from being moved from field to field. Moving animals around is not compulsory. For example, now that we have sheep, we give them free range over all the fields. But when we had ewes, tups and cows, we had to keep them separated, so they were moved around. If you are going to be moving animals, you want to plan the layout to make it as easy as possible. Easiest is where the fields all interconnect. At the other end is moving stock along roads. That can be pretty stressful.

What follows are some basic infrastructure requirements for smallholdings.

Fencing. First, there are your borders to look at. The last thing you want is your animals escaping next door or vice versa. It is not unknown for animals to jump fences and breed with the tup or bull next door. With border fences, you need to establish ownership before effecting changes or repairs. Nevertheless, it is worth checking them thoroughly. Where we are now, our borders are largely marked by drystone dykes. Where these have been damaged, fencing has been erected. One of my tasks has been restoring the stone dykes. Fencing only lasts a few years, whereas stone dykes can last for over a hundred years, so it makes sense.

Second, there may be internal fencing and gates marking out paddocks. Bearing in mind the kinds of animals you wish to keep, have a think about the layout and whether you want to change things. Ask yourself whether the fences are in the right place. It is a lot easier to re-arrange things before the animals arrive.

Third, there are fences to keep things out. The most likely of these is a fence around the vegetable patch. Vegetables are a magnet for chickens, rabbits and deer. The kind of fence you need will depend on where you are. It is worth checking out what your neighbours do before splashing out on six-foot deer fences.

One other thing, different animals require different types of fence. The standard stock fence will keep most sheep in, but not cows or pigs. Cows need barbed wire strips along the top; pigs need barbed wire strips along the bottom.

If you are thinking about electric fencing, that is covered in the chapter on equipment.

Vegetables. Chances are there may already be a

vegetable plot in place. It is worth taking a step back and checking its location is suitable. It is also worth checking it is big enough. In both smallholdings we have lived in, we have had to scale up the existing patch to about four times the original size.

It is also worth thinking about fruit. Fruits like raspberries, gooseberries and blackcurrants can be grown pretty much anywhere. As perennials, they work best in a permanent home. You may also want to have a fruit cage. Blackbirds can strip a fruit bush bare in a matter of minutes.

Depending on the climate where you are, you may want to think about a polytunnel. To be honest, polytunnels can be useful all over Britain. A lot will depend on what you want to grow. The bigger the better too. We never seem to have enough space when it is time to pot up plants from seed trays. If you are putting up a polytunnel, it will pay dividends to think about the weather. It can be pretty blowy on the west coast and on the Isles of Scotland in particular, so it may need to be strengthened in these areas. Also, further north, plan for large snowfalls; these can collapse a polytunnel if you are not careful.

Buildings. The main outdoor building you will need is a secure shed to store valuable equipment. Rural crime is, sadly, on the rise in the UK. It is nigh on impossible to get insurance unless your valuable equipment is locked away.

Aside from that, the kind of buildings you need will depend on your plans. If you are keeping animals, you will need to think about field shelters and winter quarters. Knowing your ground is important for field shelters. You don't want to be building those where the ground gets wet. Aside from becoming muddy, either the animals will end up lying on a cold, wet surface or you will be facing a huge bill for straw.

For pigs, you can get pig arcs. These are fairly simple to build. It is getting the curved tin roof sheets that can be tricky – and by tricky, I mean expensive.

Unless you have hardy cows, they will need winter housing. Cows in sheds will need to be fed and watered daily. The sheds will need to be mucked out regularly. All in all, custom-built cowsheds are the best bet if you want an easier life.

For sheep, in our experience, field shelters are very important. Maybe less so for hardy mountain breeds, but with our weather seemingly getting more extreme, the chances of prolonged heavy rain are increasing, and no creature likes to be out in that for very long.

Chickens are fairly straightforward. We have plastic chicken coops with automated door openers. The plastic coops are much easier to clean than the wooden ones, which is important given the prevalence of red mites. The automated door openers are not vital, but the further north you are, the earlier the sun rises in the summer. Chickens hate to be cooped up in daylight and, if kept in during the day, can even attack each other. So, if you are not planning to go to bed late and get up very early, automated door openers are brilliant.

The other thing to think about is whether you plan to breed livestock. Most livestock need somewhere dry and safe to give birth. Some hardy breeds of sheep and cattle are best left outdoors. Highland cows, for example, like to hide their calves amongst wild vegetation when they are born. But for the rest, a shed with light and power is essential. If your animal needs an emergency caesarean, you want the vet to be able to see what they are doing.

With all of these buildings, the key to an easier life is location. In our current holding, we arrived believing it to be set up for sheep. But after a few weeks, we realised everything was in the wrong place. The large sheds were inaccessible to the animals, and the sheep-handling area was in the middle of nowhere. The lambing area was a 600m walk – not something that is much fun at 2 a.m. in freezing rain or icy winds.

Our guiding principle is to keep the animals as close to the house as possible. This has many benefits. The first is that it helps to keep predators away. Badgers aside, most predators such as foxes will keep away from human dwellings, especially if you have dogs. Crows, ravens, foxes, badgers and even neighbours' dogs all pose threats to lambs. If your lambing shed is near the house, they are more likely to keep away.

Another benefit is that you don't have to go far to carry out your checks. During lambing, we check on our ewes every two hours. This is a lot easier when they are just a few yards from the house rather than half a mile away. It is closer for the vet if they have to be called out and a lot less walking when you forget something you need. Plus, it is nice to able to look out the window and see your animals close by.

It is also worth thinking about handling areas. By a handling area, I mean a place set up for checking animals and giving them medication. Each type of animal requires specific handling. Some of these areas can be temporary, for example using hurdles. Others need to be more permanent. In the chapters on animals, I discuss in detail the specific types of handling areas needed for each type of animal. I also discuss our philosophy of gaining trust to help manage animals. It is much easier if you can get them to come to you and also to walk into a handling area themselves.

Finally, there is water. Plants and animals need water. Carting water around is a laborious process. If you don't already have water troughs in your pasture, you may need to install these and connect them to either a mains water supply or a rainwater harvesting system.

The water systems we installed capture both rainfall and groundwater into tanks, and these are connected with MDPE pipe to bowsers in each field. It took a bit of planning, but it works well. Two tanks fill up from groundwater streams, two tanks collect rainwater from the sheds and the bowsers refill automatically from these.

Most of what I have mentioned here can be built by yourself. You can get premanufactured sheds of all shapes and sizes that you can assemble yourself. Obviously, the bigger they are, the more useful it is to have help. Fencing is pretty straightforward, it just takes a lot of muscle power. You can get a handheld post stomper from most country stores. You can also get tractor attachments for knocking in posts. These work well in flat areas but can be a little tricky on hillsides. Modern plumbing is also straightforward – you can have push-fit or pressure connections that can simply be hand tightened.

There are also plenty of contractors who can build field shelters, install fencing, and so on. The choice is yours. We pretty much do it all ourselves. For example, we hired a man and a digger to dig a hole for our large water tank but then did the rest ourselves.

So, in summary, a little forethought in planning your infrastructure can go a long way to making your life easier. There is always a lot to do in this life, so automating things can save you a lot of time.

11

BEES

"Let's get bees," we said. Nicole likes honey, and we both like bees. Some of my friends in the past had kept bees in their suburban back gardens. And there is much talk about how bees are struggling these days. We had the space. How hard could it be? We bought a book, read up on keeping bees and took the plunge.

We ordered a hive and all the relevant accessories to go with it. We chose the hive to match the chicken runs. It was made of plastic, unlike the traditional wooden hives, but all of the literature and reviews made it out to be an excellent product. It was also touted to be much easier to use than a standard hive. To be honest, it did look pretty good once unpacked and assembled.

We ordered a nucleus of Buckfast bees. We chose Buckfast bees as they have a reputation for being hardworking yet chilled out at the same time. Being novices, we wanted to get bees that were easy to handle. The bees duly arrived in the post delivered by a mildly concerned postman wondering what the buzzing noise was. He was pretty surprised to find out he'd had a swarm of bees in a box in his van.

We introduced the bees to the hive and retired to a safe distance to let them get on with it. While we had not joined our local beekeeping association at that point, we did find a local expert who was happy to be our mentor. He had been keeping bees for 75 years, so he knew a thing or two. The puzzled silence coming from him as he viewed our plastic hive was a bit worrying, but we put it down to it being a new thing and him being a little traditional. We opened up the hives and *the bees seemed pretty happy, so all seemed well.*

They were great bees. They were so chilled that you could stand in front of the hive and they'd just fly around you to get in and out. Even cutting the grass didn't faze them. Bees are not keen on vibrations, but the mower seemed OK; I was able to pass the front of the hive mowing the grass with no protective clothing. They were lovely, and I really looked forward to opening the hive and checking in on them.

I did get stung though. I wasn't even anywhere near the hive or the bees. I was just standing in the garden minding my own business when one flew into me and stung me on the cheek. That's when I found out that I had become slightly allergic to bee stings. It was the day before our wedding, and my cheek swelled up nicely in time for the ceremony.

Anyway, the bees settled in. As it was their first year and they needed time to build up the colony strength, we decided not to take any honey. As the weather cooled, over a number of evenings, we noticed a few bees seeming to struggle to get back into the hive. They were arriving back laden with pollen but getting stuck at the entrance. There was a little ridge just inside which they were struggling to crawl up. I suspected they were cold, so we gathered them, one by one, into the palms of our hands and gently breathed warm air onto them. After a few seconds of this, they would liven up a little and were then able to crawl back into the hive.

When winter came, it was a pretty wet one. There were puddles everywhere pretty much the whole time. As advised, we left the hive alone so as not to disturb the bees during the cold spell. The literature coming with this hive said it was well insulated. But, sadly, when we opened the hive to check on a sunny day in late winter, we found that the bees had died.

We don't know why this happened. We consulted our mentor, and he could not see a reason either. Perhaps the wet winter had caused condensation in the hive. There was also a problem with beekeeping generally across Europe – the varroa mite. But we never found out why ours had died. It was quite distressing really.

Our mentor clicked into action and found us a new nucleus at an apiary not far away. He had done well as it was slightly off-season for acquiring nuclei. I duly drove over and collected the new nucleus. These were not Buckfast, but we were assured they would be just as gentle.

They turned out to be anything but!

Within days of them being installed in the hive, we experienced problems. These bees were very aggressive. In fact, if you looked through the kitchen window, you could see bees strategically positioned all around the garden. They were like sentries. It felt like they were just waiting for us. The thing was, they were; the moment we went out, they would zero in on us and fly at our heads in an angry rage. It was horror film material.

Getting to the parked cars became stressful. Having got there, seconds of panic ensued while you checked no bees had followed you into the car. Gardening became a perilous activity. In fact, we had to wear bee protective clothing just to go out. As we crouched, weeding, you could see two bees peeling off the top of the shed and flying in like a pair of fighter aircraft. There was no escape.

We called our mentor. He came, had a look and identified the problem. The apiary had allowed the bees access to the feeding area of the nucleus box and they had laid brood there. When we had put the frames in the hive, the brood in this feeding area had not been transferred. The bees had accidentally been split into two swarms, one queenless but with brood to protect. He sorted it out and then rang the apiary to given them a right telling off. He was furious. While it was something to behold, I got the impression that the apiary didn't really care. A sad reflection on modern customer service. We checked online and sure enough, they'd had a few bad reviews.

Things settled down, a bit. The fighter aircraft behaviour stopped, but the bees never seemed to settle. They just seemed to be angry all the time. Checking the hive became a pretty scary activity. You certainly made sure you had closed every zip and overlapped every bit of Velcro properly on the bee suit. There was no way you would go near the hive otherwise. Even protected, having hundreds of angry bees swarming incessantly around your head is unnerving. They even tried to sting me through the suit.

So, the checks became more superficial and the swarming season arrived. To be honest, when the first swarm headed off, I was secretly relieved. Bye-bye angry queen, I thought to myself. Nevertheless, with the help of a local beekeeper, we gathered the swarm. He took it away so that it could be rehomed. In fact, they swarmed three more times. Each swarm was captured and put in a hive. We were getting low on bees, so one swarm was merged back into the main hive under the watchful eye of our mentor. In fact, word had got around that I kept bees; so I ended up capturing a further two swarms in other gardens. I don't think they came from our hive, but you never know. We rehomed a couple and ended up with a second hive further down the garden.

It wasn't long before this second hive started exhibiting problems. I was getting the hang of bees now and sensed something was not right. I asked our mentor to take a look. The hive was queenless. Our mentor sourced a new queen and split the hive in two. He showed me how to introduce the queen. The plan was to get her established and then merge the two hives back into one.

Then the robber bees came. There were bees everywhere. They were in the shed, they were in the trees, they were flying around all over the place. Strangely, they were not aggressive. You could walk amongst them, but they just ignored you. It was a little unnerving to say the least. But the upshot was that they raided the new hive and killed the new queen.

It was quite depressing really. There was just one problem after another. This beekeeping was turning out to be much harder than anticipated. I took the executive decision to merge all our bees back into a single hive. Our mentor told us what to do and over a few days, the hives were merged. Thankfully, this worked; and with swarming season over, things settled down.

In the meantime, some friends in a nearby village asked if they could have some bees. The first swarm that was with the other beekeeper was still available, so I arranged to collect it to take it over. I turned up to make it ready only to find that I had forgotten my gloves. My mentor never used gloves, so, I thought to myself, I won't need them. Wrong. The hive to be collected was at the end of a long line of hives and before I got halfway there, I had been stung on the hand. Mildly embarrassed, I borrowed a pair of marigolds.

The transfer all went fine, but the bee sting affected me. In fact, I had a range of symptoms. When I entered these into NHS Direct, its advice was to call an ambulance. I didn't as I wasn't that bad. But it was clear that I had become allergic to bee stings.

That winter, we moved our hive to the furthest corner of our field. The love affair with the bees was wearing off. Even then, with a whole orchard of apple trees and a new queen, the bees were still as angry as ever. After inspecting them, they would follow me for up to 400 yards before leaving me alone. With the fear of being stung looming over me, and probably making me nervous around the bees, we decided to offer them for rehoming. They were soon snapped up by another beekeeper.

We never did get any honey.

What we did learn is that having a mentor is vital. With all the books and training courses you can have, when things go wrong you need expert help and fast. We are eternally grateful to our mentor, Ivan. An amazing man who really knew his bees.

12

EQUIPMENT

Running a smallholding does require some equipment. There is plenty to choose from. With the increase in mechanisation and automation across farming in general, equipment of all shapes and sizes have sprung up, all purporting to make your life easier. It is all too easy to be seduced and buy a whole lot of stuff you don't really need. It is a bit of a catch-22 to be honest. Until you have lived the life for a while, you don't really know what you need. But you need the equipment while you are leading the life. Buying what you don't need is an easy trap to fall into. We have bought more than one piece of equipment that seemed vital, which then sat in the shed for a couple of years and was sold unused.

When we first started, we grew vegetables and kept chickens. The only equipment we needed for the vegetables was netting. Birds and butterflies can cause a lot of damage. If you have spread a lovely layer of mulch, then the blackbirds will have endless fun digging up the mulch and your carefully planted vegetables while looking for worms. The cabbage white butterfly can decimate your brassicas. We tried buying netting frames, but these were generally useless insofar as they were too fragile and broke too easily. We now use lengths of plastic pipe supported by wooden posts. They work perfectly.

We spent a lot of money on chicken equipment, though.

I had already lost two flocks of chickens to foxes in my town dwelling, so, even though country foxes are much rarer than their town counterparts, we were determined that our country chickens would not meet the same fate. On moving to our first smallholding, we purchased a large, enclosed fox-proof run. These are available online and are expensive. Our run comprised a good-sized walk-in run onto which I added a custom-built extension. The chicken coop door was left open permanently so they had access to the fox-proof run at all times. This meant they could get up and go to bed as and when they pleased in complete safety.

This kept the foxes at bay, but even though the chickens had way more than the recommended one square metre each, there was still a lot of bullying and the ground in the run was churned up into quite a mess.

We tried to alleviate this by letting them out of the run into the garden and paddock so they could range freely. We could only do this when we were around to keep an eye on things. As we both worked, it tended to be mainly at weekends. They did enjoy their excursions out, but it also meant that we had to put up quite a few fences to protect our flower beds and vegetable patch.

We also had problems with red mites, and it proved impossible to clear them from the wooden chicken house. They got into all the nooks and crannies, and it couldn't be disassembled to clean out those areas.

In the end, we burnt our wooden coops and replaced them with plastic coops that had automatic doors. This worked out better for the hens and us. The doors open at dawn and close at dusk. We also have rainproof automatic feeders. Most of the chicken feeders for sale have roofs that are too small, and the feed gets wet and mouldy. The best feeders are those where the chicken stands on a platform and that opens access to the food.

One of the questions we are often asked by people thinking of taking on a smallholding is whether they need a tractor or not. It is a good question. The answer is probably not. However, there are certain activities made much easier with a tractor. Tractors come in all sizes, and you can get quite a good compact tractor second-hand for a reasonable price. If you do so, make sure it has the standard A-frame at the rear for attachments.

We have a tractor. We ended up getting one when we bought sheep. We had a large expanse of grass, but we just couldn't keep it under control even with a ride-on mower. So, we got sheep. But in the fast-growing summer months, the sheep couldn't keep up and the grass grew too long for them. Sheep like to nibble. So, ironically, we had to buy a tractor and topper to keep the grass down for the sheep.

However, it has proved useful. We have a fairly long track to maintain, and the front loader is great for fetching scalpings with which to fill in the potholes. We now buy large bale hay, and the tractor is great for moving these around. Also, a tractor is most useful for pulling large bowsers full of water into the fields during long dry spells. That said, most of these tasks could be carried out using a quad bike or decent 4x4 and trailer, so a tractor is not essential. But it is something to think about.

While we are on the subject of tractors, there is a wide range of implements available. You can get hedge cutters, front loaders, backhoes, ploughs, harrows, toppers, log choppers, post stompers. The list goes on. Not all of these fit all tractors. If you have a list of what you need, make sure you can fit them to your tractor.

That brings us on to trailers. If you have livestock, you are likely to need a trailer. When you buy animals, you are expected to collect them. You may also be interested in taking them to shows. It is quite hard to find people who can move small numbers of animals around. The size of trailer will depend on the numbers and types of animals you have. We also find a flatbed trailer invaluable. We didn't get one for years, then, when we did, we wondered how we'd managed without it. There is quite a good market in second-hand trailers. There is a wide choice of trailers, so it is worth spending some time thinking what you want to do before heading out to buy one. Otherwise, like us, you will soon be selling the trailer you've just bought to buy the one you should have bought in the first place.

For us, our most useful piece of equipment is the 4x4 quad bike. It is invaluable for zipping around; we use it for a wide range of jobs. From taking winter feed out to the feeding stations, to ferrying sheep poo and straw to the compost heap. From ferrying mulch to the veggies, to carting bags of weeds. It is probably our most essential piece of equipment. If you get one, just make sure you lock it away when not in use. It is the number one item targeted in rural crime.

If you are considering making your own hay and/or growing arable crops, you will need to think about harvesting. Haymaking requires a cutter, haybob (for turning the hay) and a baler. This is an expensive set of equipment. We priced it up, compared it with our annual bill for hay and worked out it would take 15 years to pay for itself. We decided to carry on buying winter feed.

This level of equipment only really pays for itself if you can get contract work. One of our neighbours does exactly that: he cuts and bales hay for small farms in the area. It is a thought as there is a continual demand for small bale hay yet not much of it around. So it is a way of making money. Just beware, the machinery is complex and prone to breaking down. Ideally, you need to be able to repair it yourself. There's nothing more soul destroying than cutting hay, drying it, then watching the rain ruin it as you wait for a mechanic to fix your baler.

That said, you can always return to traditional methods and use a scythe. Much cheaper and, according to people I know who use this method, quite rewarding. There's a lot to be said for traditional methods. Some of these skills have drifted out of use, but there are many courses across the country where you can learn. Tractors effectively replaced horses, so, on a small scale, using a horse and cart could work pretty well.

Livestock need to be inspected and vaccinated. Again, there is a huge amount of equipment on offer that claims to make things easy. Each animal requires different equipment, and there are some essential items.

For sheep, sheep hurdles are a must-have. You can create temporary pens of any shape and size at any location with these. For cows, you also need hurdles. However, these are much larger and heavier and really quite difficult to move around.

Cows need a permanent handling area with, at the minimum, hurdles and a yoke. You really need to immobilise a cow before giving it an injection.

Pigs need a pig arc. They also need to be well fenced in as they are expert escape artists. It is common practice to use an electric fence. These work, given you have trained the pigs (more about this in the pig chapter), but make sure the battery does not run low. They will know. Also, giving a pig an injection is pretty challenging. There are things called pig boards which are supposed to enable you to trap the pig in a corner. Let's just say it is not something we managed to master.

There is also a plethora of small stuff you need, mainly to do with ear tags and injections.

As mentioned above, these are detailed for each animal in the relevant chapter. It is worth costing what you need before you buy the animals. Otherwise, you could find yourself with a hefty bill for equipment you didn't realise you needed.

13

WITCHY

Our first lambing was over. We had been through weeks and weeks of interrupted nights and, therefore, pretty tired. We had experienced real highs and lows. The experience of delivering your first lamb is hard to beat. Burying a dead lamb is something you really never want to do again.

It was Easter, Good Friday to be exact. All the sheep and lambs were out in the field. It was a bright sunny day. I was getting ready to head off to work. Then I noticed something strange. We had put George, our Anatolian Shepherd dog, in with the sheep to guard them. The sheep know George and were happy to have him around. Well, normally they were, but right now, Selene was chasing George around the paddock.

I went down to investigate and let George out of the paddock. He was very thankful to escape, let me tell you. I went over to Selene only to find that her waters had broken. This was something of a surprise as we had thought Selene was not pregnant. In fact, she had been turfed out in the field with the early lambers so she would not have access to the sheep nuts. Not being pregnant, she shouldn't really have sheep nuts.

Yet, here she was about to give birth. I fetched Nicole and we sat down with Selene. After I rang work and explained what was happening, they told me not to come in. That was great.

In no time at all, Selene popped out two lambs, a boy and a girl. We named them Warlock and Witchy. All in all, the lambs seemed perfectly healthy. Warlock was straight in and suckling for all he was worth. Witchy was also up and about but did look a bit thin. Selene, no longer carrying unborn lambs, suddenly looked a little on the thin side, too. She had needed those nuts after all.

It was sunny and warm and there was plenty of new grass. Selene was off grazing with her new lambs in tow. On the surface, all seemed well. However, over time we noticed that Witchy never seemed to get a chance to suckle. Every time she got near a teat, Selene would move on. It was not that Selene was a bad mother; she was just hungry. While we follow the motto that mother knows best, sometimes you have to intervene. The thing is, it can be hard to know exactly when to intervene. On top of that, a field birth is actually less risky than a shed birth. All had gone well. Selene was a proven mother. But there was something about Witchy's behaviour that made us hang around and keep an eye on her.

After a while, we saw Witchy wandering off and lying down. Not in itself unusual, but this reminded of us of how Larry had behaved when he had given up. We suspected Witchy was not getting enough milk, so we decided to intervene. We brought Witchy indoors and put Selene and Warlock into a lambing pen. Warlock was doing well and seemed to be getting plenty of milk. However, penning them up would make it easier for Warlock, plus make it easier for us to keep an eye on things, just in case.

As mentioned, Selene had not had the required supplement of sheep nuts during her pregnancy and she was a bit on the thin side. She was also now providing milk, so we gave her sheep nuts along with her hay. The mistake we made was to give her too many too quickly. Within hours, she developed acidosis. She nearly died. That was a real wake-up call for us. Sheep nuts have their place in a sheep's diet, but they need to be handled carefully. Too many too soon and you could have a dead sheep on your hands.

Meanwhile, Witchy was now safely indoors and warming up. We prepared some colostrum and offered it in a bottle to Witchy. We reckoned she must be pretty hungry and would welcome something to eat. Witchy had other ideas. We tried and tried to get her to drink from the bottle, but she was not having it. In the end, we had to tube feed her. This is something we only do as a last resort, but Witchy was *losing her strength and we had to act.*

Nicole sent me to bed to catch up on sleep while she stayed up with Witchy to try and get her to bottle-feed. At 4 a.m., Witchy started to go downhill. Nicole opened the laptop and looked up how to give lambs an injection. Nicole suspected Witchy might have watery mouth from a lack of colostrum and stress, so she wanted to give her an antibiotic.

Nicole had never done this before; it was really stressful. However, she followed the instructions on YouTube. With Witchy in her lap, she found a teeny bit of muscle on her skinny leg and administered the injection with trembling hands. She then kept Witchy warm and waited. It was the most nerve-racking experience in Nicole's life. About two hours later (it is all still a bit of a blur), Witchy started to wriggle around a bit. Nicole made up some electrolyte and popped the teat in her mouth. Witchy drank for all she was worth. It was more than Nicole could have wished for.

Nicole staggered upstairs to bed with Witchy and popped her in a box next to her and tried to grab some sleep. About an hour later, we heard noises; it was Witchy moving around in her box. It was the best alarm call you could have wished for. We gave her some milk, and this time she took the bottle readily and guzzled away for all she was worth. After that, to keep her warm, we put her in bed between us. Later, after another good feed, we decided to put her in with Larry and Lisa (rather than place her back with Selene) just for a while to get her strength up.

It took about 36 hours for Witchy to get back to full fitness. Two nights and a day had passed. We spent some time discussing whether to put Witchy back with Selene. We were a bit worried that Selene might have forgotten Witchy. Also, Witchy would be covered in strange smells. The ewe may not recognise it if the lamb does not smell right, Mothering ewes always smell the lambs coming to suckle to check that they are theirs. If the lamb smells wrong, they are sent packing. With Selene and Warlock in a lambing pen, there would be a risk of Witchy getting injured were Selene to reject her. It was a dilemma.

Sunday morning came, but we still hadn't decided what to do. Nicole took all three orphans out. By now, Larry and Lisa were going out every day, so we were just allowing Witchy to tag along. At this point I was still inside the house when I heard this almighty "meh". Even through the walls it was loud. It was Selene. She had spotted Witchy walking past and was calling out to her. Witchy answered with a mighty "meh" of her own. They had remembered each other; this was great news. Nicole quickly took Witchy to Selene. However, as we feared, Selene did not recognise Witchy's smell and would not let her suckle.

Nicole took Witchy out and into the field. By now I had joined them. We discussed what to do and decided to rub some of Selene's milk over Witchy and try again. This worked a treat. Mother and daughter and brother were back together in no time. We were ecstatic. We were already learning that the best place for a lamb is to be with their mother. Witchy was back where she needed to be. And now she was strong enough to suckle.

They were, all three, soon allowed back out into the field with the others.

In our first year, we sold most of our lambs. But Witchy is still with us – we kept her as we couldn't part with her. She has blossomed into a strong, healthy sheep and is now a mother herself. She gave us Yogi, our cute ginger lamb.

14

GROWING FRUIT AND VEGETABLES

Growing your own fruit and vegetables can be hugely rewarding. It is true that they taste better. This is especially true of fruit. Many supermarket varieties are bred for longevity and shape, but this is often to the detriment of taste. Your own produce can be eaten fresh from harvesting, and that makes a real difference. Also, supermarket vegetables tend to be grown with the use of pesticides and herbicides, some of which are absorbed into the plant. Grow your own, and you can be confident they are chemical free.

There are many excellent books on how to grow fruit and vegetables. They cover important subjects such as rotation and companion planting.

The focus of this book is on the balance of time, effort and money that it takes to run a smallholding. In this respect, growing vegetables is not too different from keeping livestock. No matter what you might think, vegetables don't just grow by themselves. They need a lot of care. It is an interesting fact but, if you have a new vegetable patch, the first year can result in bumper crops. However, by the second year, the pests have found it and can cause a lot of damage.

Starting at the beginning, the first thing you need is an area in which to grow your fruit and vegetables. There will likely be one area for fruit bushes (perennials) and another for the vegetables (annuals). Herbs kind of fit into both, so it's normal also to have a separate herb garden. Now, you may be lucky and move to a smallholding with the perfect sized vegetable patch that has been well cared for over a number of years. The fruit bushes look healthy and well pruned, and the herbs likewise. Brilliant!

The thing is, that has never happened to me. Every allotment I have taken over has needed sorting out. Every garden I have owned has needed much attention. The two smallholdings Nicole and I have taken on have both needed major work carried out on the vegetable and fruit growing areas.

So, it is likely that you will need to spend time digging, either creating a new area or extending and repairing an existing one. If you are prepared to wait, you can always lay a membrane over the weeds, but this takes about a year – a whole year of lost production. Even then, certain weeds just wait there, dormant, until you lift the membrane, and then they return with a vengeance.

Digging is at least free, but it is time consuming. On top of that, you may find yourself laying paths or even building raised beds. It is easy to spend a lot of money buying slabs, sand, concrete and wood when setting up a vegetable garden.

There is also the question of compost. Have you inherited a workable compost system or will you have to set one up? You can buy ready-made compost bins which can be assembled quite quickly. Or you can build your own. Again, it's a question of time and money. Making compost is really worth it, and it's also one of those few activities which does not take too much time or money. It just helps to turn and aerate it occasionally. That said, if you are collecting livestock manure, that can be hard work.

Of course, you could buy in mulch and compost, but it is amazing just how small an area a large bag of compost covers. Buying mulch and compost can soon prove expensive.

Vegetables

Most vegetables are annuals so need to be planted from new each year. Seeds and seed trays are relatively cheap. However, the sowing of seeds takes quite a bit of time, especially if you do it properly.

Having sown your seeds, the next question is where are you going to put them? While some seeds are hardy and can cope with frosts, others are not. Some, such as tomatoes, need to be kept warm.

The best environment is a greenhouse or polytunnel. These become more essential the further north you go. However, that's not all. Seed trays take up space and if, like us, you are growing a fair few vegetables, then you'll have a lot of seed trays. That means some sort of shelving system that provides sufficient light. You may also need some heating.

Again, some of this may be in place when you move in. If not, you'll need to think through what you want (the bigger the better), and then purchase and construct what you need. Where we are now we have a greenhouse, but we could do with one twice the size. We have also had to install power so we can keep it frost-free and build shelves to cope with all our seed trays.

Of course, then the seeds germinate and need to be potted on. The mass of plants contained in a small seed tray suddenly expand into individual pots, so much more space is required. Hence, the bigger the greenhouse the better.

Some plants, once germinated, can be moved outside, but they may require protection from both frost and pests. Freshly planted pots are a magnet for blackbirds, and likewise baby leaves for slugs and snails.

The next stage is planting out. Seasoned vegetable growers will know that a little preparation goes a long way. Planting out your young, fragile seedlings into well-prepared soil is the best way of getting a decent crop. This sounds simple, but the preparation can be fairly complex, involving PH measurements and the use of fertilisers. That said, a good layer of mulch over the winter pays huge dividends.

Planting out is, again, fairly time consuming. However, the real issue now is pests. Once your carefully nurtured baby plants are out in the open, they are vulnerable.

Pests come in many shapes and forms. Some plants have their own individual nemeses such as carrots and carrot fly. But the worst three, in our experience, are birds, slugs and butterflies.

Specialist vegetable growing books talk about these pests and approaches to dealing with them at length. With so many strategies suggested, it can be hard to choose what to do. The problem is, strategies for dealing with one type of pest can work against dealing with others. For example, putting up bird netting then keeps out gardeners' friends including hedgehogs, frogs and toads.

Birds are a pain but, as a rule, they don't eat your crops. Sparrows can be a bit partial to young salad seedlings, and pigeons quite like broccoli. But the real damage is caused by blackbirds. They like to dig around looking for worms. If you have just planted out seedlings, the blackbirds will soon be digging them up. I love blackbirds, just not in the veggie patch.

For us, the worst offender is the caterpillar of the cabbage white butterfly. You have a choice. You can put netting over to keep the butterflies off or you can pick the caterpillars off each day. We did try letting the hens in to see if they'd eat the caterpillars, but they preferred to eat the broccoli leaves, so that didn't work. Having tried both netting and picking off caterpillars, we'd definitely recommend netting.

Netting is just another expense you incur. One thing, though, we wasted a lot of money buying frames that turned out to be fragile and, therefore, basically useless. Buy the netting, but make your own frames. MDPE plastic pipe and wooden posts work well for us.

One proviso, butterfly netting can also keep out pollinators. If your crop needs pollination, you can't use butterfly netting until the pollination is complete.

So far so good. But now we have to think about gardeners' main two nemeses: the slug and the snail. What do we do about them?

The easiest solution is organic slug pellets. We are not that keen on using these, as they can be harmful to wildlife. However, we do sometimes, in desperation, resort to using them in the netted-off areas. The nets mean that birds, hedgehogs and so on are not exposed to them. Another problem is that if you have netting, it is likely to be held down by bricks or some other type of heavy object. These provide the perfect hiding place for slugs. So using netting can inadvertently lead to an increase in slug damage.

Another approach is to build specially designed areas for different crops. For example, carrots can be grown in a sandbox with tubes of compost. Potatoes can be grown in boxes of sterile (bought in) compost. There are numerous ideas out there for building environments that act as pest deterrents.

It's a balancing act. You can choose to set traps for slugs and snails, but that means they must be inspected regularly. You can leave the netting off, but you'll need to pick caterpillars off at least twice a day. You can build special areas for certain crops. Once again, you face the balancing act of choosing to spend money to save time or vice versa.

Overall, the first year can prove quite expensive. There may be a lot of equipment to buy and set up. The good news is that most of it is reusable.

After that, it's just finding the time given all the other things you need to do. It is oh so easy to forget the vegetable patch.

Fruit

Fruit comes on bushes or trees. Trees pretty much take care of themselves. If you are planning to let livestock graze around fruit trees, you will need to protect the trees; otherwise, the livestock will help themselves to both leaves and bark. You could end up with no fruit and no trees.

The focus here is on fruit such as strawberries, berries and currants.

Strawberries aside, these tend to grow on prickly bushes. All are favourite foods of birds. Fruit tends to ripen in late summer, so the birds will be up and stripping the bushes long before your alarm clock goes off. You have a choice: grow enough for all of you, birds included, or protect the fruit.

You may be lucky and have inherited a fruit cage. If not, you may need to build one. Again, you have the choice of building our own or purchasing one. Time and money, again.

If you are setting up a new fruit cage, the most important things to remember are that you need to be able to get in and around the bushes. It also helps to have wheelbarrow access. So make it as tall and spacious as you can.

We also plant a few bushes outside just for the birds. It seems only fair.

Harvesting

One final thing to consider is harvesting. When I was growing up, you bought vegetables at greengrocers. Most were bought loose and looked like they had just come straight out of the ground. Root crops such as potatoes and carrots were still covered in soil.

All vegetables had to be prepared. An activity involving any or all of the washing, peeling and chopping. Salad crops needed to be inspected for slugs and caterpillars. This time-consuming operation was often delegated to us children.

These days, most vegetables come shrink-wrapped and ready prepared.

Harvesting your own vegetables is great, but it takes time to collect them and more time to prepare them. While the end result in terms of taste and texture is worth it, it is another time pressure to fit into a busy lifestyle.

The temptation to keep a shop-bought packet of frozen peas in the freezer can be hard to resist.

15

LUCKY THIRTEEN

Our hens and cockerel (left behind by the previous house owners) have a large area in which they can roam. At that time, the run was bordered by hen netting alongside a hedge on one side and a drystone dyke on the other. It was not entirely hen proof; it was more to keep dogs out than hens in. However, we noticed that they rarely left the run. They must have been happy with the space they had.

One day, Mrs Mills (one of our Maran hens) just vanished. We have electric door controls that open the chicken coop door at dawn and shut it at dusk, so we thought it unlikely a fox had taken her. Also, it would be pretty unheard of for a fox to take just one chicken from a coop. We do have buzzards and kites around, but, again, it would be unlikely they could take a fully-grown hen. Hens can be pretty fierce.

We kept an eye out, but there was no sign of her. Then, a couple of weeks later, she suddenly appeared on the front lawn. She looked a bit dishevelled and moody. We knew straight away – she was broody. We gave her some extra food and then hid so we could see where she went. She was very wary, but eventually she carefully sneaked back to her nest. We managed to follow her and found that she had built a nest under a bush and some bracken just metres from the house.

It was a worry that she was exposed to foxes. She was also exposed to some pretty harsh weather. We had some heavy rain that month, and bracken can only afford so much protection. But, at the end of the day, what can you do? We had to trust that Mrs Mills knew best, so we left her to it. When she popped out for some food and water, we fed her some high-energy scraps just to help her along a bit.

We did manage to sneak a look at the nest while she was out and about, and we counted 13 eggs. Given Mrs Mills' mother, Hattie, had sat on a lot of eggs but only hatched two, we were not sure what to expect.

One warm, sunny spring day a few weeks later, Mrs Mills finally emerged with her chicks. She had hatched all 13. We were amazed. We sat out on the grass, and the chicks wandered all around us and even over us. We put down some tiny pieces of chopped up strawberries for them, which they loved. They were so light they could climb tall daisy stems. They were great.

When tired, they'd all pile under Mrs Mills until you could see bits of chick popping out all over the place. A head here, a wing there, a foot under there. Always moving. Mrs Mills looked really chuffed.

For a few days, they foraged around the garden area. Then, one day, Mrs Mills decided to take them into the hen run. That was easier said than done. She tried to lead them over a stone dyke. Amazingly, one or two managed to follow her, but the rest were running around the base of the wall on the wrong side. We sprang into action. Between us, we managed to catch the stragglers and put them into the run. Calm descended once more.

At least, it remained calm until bedtime. At this point, Mrs Mills tried to lead them back to the nest. The wall on the hen run side is much higher, so none of the chicks were able to follow her. Mrs Mills stood on the top of the wall calling while the chicks ran around in a bit of a panic. This time, we decided not to intervene. To be honest, we were hoping that Mrs Mills would move into one of the chicken coops so we could keep them safe at night. Mrs Mills had other ideas. Having had a taste of wild camping, she was not about to go somewhere warm and dry.

She eventually chose to nest under a single-stemmed loganberry bush inside the hen run, next to the wall. There were a handful of nettles to give extra protection, but it was pretty exposed.

We were not sure what to do, so we left her out for the night. Thankfully, nothing happened. It may be that the smell of our dogs kept any nearby foxes away. Whatever it was, all 13 chicks were still going well. But we still wanted Mrs Mills in a chicken coop.

We came to the conclusion that we had to engineer things so Mrs Mills would choose the coop. We felt if we just picked them all up and put them in there, the next day she could revert back to wild camping.

We waited until near bedtime, and then set up some temporary netting around both the loganberry bush and the coop. Luckily, her nest was quite close. The next stage of the plan was to get Mrs Mills and her brood of 13 into the netted-off area. Well, that was interesting. We could get a few in, but some of the chicks were still small enough to get out through the netting. In fact, we had to wait until they were ready to go to bed before we could get them all in.

At that point, we cut down the nettles and loganberry leaving a bare patch. We settled back to watch. Mrs Mills was most put out. She did keep settling down there (where the bush had been), but then seemed to realise that she was exposed so got up and walked around. It took a fair time, but eventually she looked inside the coop. Our inner cheers were stifled when she came straight out again. It probably took about five visits over half an hour before she finally led the chicks in and settled down for the night.

Job done, or so we thought. In fact, we had to repeat this process for three days. After that, Mrs Mills finally chose to use the coop at night, so we were able to sleep better, too. Also, they were all out of the rain (it rains a lot here).

As the chicks grew, they became more and more independent. This created a whole new challenge. While still unable to fly and therefore unable to get on top of the wall, they soon discovered that they could get through the hen netting. Beyond the hen netting was a large area of grass and then the main track up to the house. There was no stopping them – they were through the netting and off in all directions.

It was a constant worry. There were three resident cats in the world beyond the hen run. There were also cars and vans using the track. Nevertheless, the chicks came and went without mishap. How they managed that, we don't know. Bit by bit they grew, until one day they started to find that they could no longer get through the netting.

That was a relief which lasted about five minutes. Thwarted by the netting, they turned their attention back to the wall. In no time, all 13 had worked out how to fly to the top of the wall and head off yonder.

In the end, we just left them to it. Since then, the netting has been replaced by a combination of stock fence and chicken wire. Future broods will not face the same risk. We did put in hedgehog tunnels with little "doors" that the hedgehogs can push through but should be enough to foil chicks.

All 13 chicks continued to thrive and grow. By now, we were able to identify which sex they were. The males grew longer legs and you could see a nodule that would turn into a spur on each leg. The females were smaller and dumpier with shorter legs. We counted nine males and four females.

We suspected this might be a problem, but we waited to see how things would turn out. For a while, all was fine. Then one day, the males grew up. In a way, it was a bit like they had become the hen equivalent of teenagers. Overnight, they developed a more aggressive attitude. They formed two gangs and started roaming the run with intent. They harassed the hens, jumping on them and pinning them down. This was hen gang rape. It was not good to see. If this were allowed to continue, they would kill the hens. Not deliberately, but the hens could not cope with this level of harassment. Ideally, each cockerel should have a minimum of four or five hens each. We now had ten cockerels and nine hens.

This is the dilemma at the core of this life. If you breed animals, the question "What do you do with the boys?" arises. They are next to impossible to sell. Maybe a prize specimen might go, but otherwise the boys don't offer anything. Their role is to protect the hens, but you only need one or two in a flock for that. Otherwise, they are just a problem.

If we did nothing, these cockerels would start killing each other and the hens. We had to act. The decision was that they went in the freezer. While it was a hard decision, they'd had a good life. We didn't set out to breed hens. We keep them for the eggs, and we have a single cockerel to help protect them. Nature takes its course.

Peace returned to the run. The four hen chicks grew into strong adult hens and one of them, the following year, produced a single chick. Thirteen one year, one the following year. As they say, don't count your chickens before they are hatched!

16

KEEPING LIVESTOCK

When visiting the countryside, there are lots of livestock farms around. In the fields, you can see animals, mostly sheep and cows, grazing contentedly or catching a nap. It all looks peaceful and tranquil. It is all too easy to conclude from this that keeping animals can't be that hard. Well, at times it is and at times it isn't.

This chapter looks at keeping livestock in general. Each of the animals mentioned also has a detailed chapter later in the book that covers specifics. Note: this book does not cover horses and other equine breeds.

First things first, keeping livestock on a smallholding is, in reality, not that different from a skilled job. The assumption here is that you are moving into this smallholding life with little or no experience. Those people who have grown up on a farm, for example, will know instinctively how to care for their animals. If you are new to this, a steep learning curve awaits you. This learning is not cheap both in terms of emotional and financial costs. As part of this learning, you will make mistakes and animals will die. It is hard, but it is part and parcel of this smallholding life. It is a good idea to invest in training so you can learn how to look after these animals.

At the risk of labouring the point, it is very different from looking after common pets such as cats or dogs. With small animal pets, when things go wrong you take them to the vet. In fact, there's not much the vet will let you do yourself. With livestock, it is completely different. It is mostly up to you. The rural vet will expect you to undertake all of the routine care including inoculations. In fact, when we first had to vaccinate our sheep, we called the vet out thinking that's what you did. The vet administered an injection while at the same time talking me through what he was doing. I wasn't quite sure why he was going into so much detail, until the moment he handed over the syringe and basically told me to get on with it. As I said, a steep learning curve.

So, bearing that in mind, you would like to keep some livestock.

A key early decision facing you is what animals you would like to keep. What you choose will very much depend on what your goals are. Animals can be anything from pets to a food supply. Some offer secondary products. For example, hens produce eggs, sheep produce wool, and cows can produce milk. At the beginning, it can be hard to get this choice right. As well as thinking about the type of animal you would like, there are also breeds to consider.

It is worth doing the research. Each breed has pros and cons. For example, hardy mountain breeds of sheep can be pretty self-sufficient, even at lambing. However, they are more like goats; and such breeds do not take kindly to being kept in small fields. In fact, these mountain breeds are all escape artists of the highest level. Other breeds such as Ryelands can be less inclined to clamber over walls, but they need a lot of attention at lambing as they have been bred to produce multiple lambs.

Some cows have been bred for specific purposes, mainly dairy or beef. Each breed has its own behaviours, and some are easier to handle than others. We kept Highland cows. These are, in the main, gentle, but you have to be comfortable with their horns. And they can be quite stubborn. Regarding pigs, they are intelligent animals, astute escape artists and they are all big. Very big. Micro pigs are a myth!

Regarding chickens, you can get young hens that are coming up to laying or you can get ex-battery hens that are a year old. The latter are great, but they can be fragile until they settle in.

Whatever your thoughts, there's no need to rush into a decision. It can sometimes be worth taking your time and planning the introduction of animals onto your patch gradually. It is down to you and how much time you have.

Time is something to think about. Animal husbandry takes time. On the face of it, animals' basic needs are food, water and shelter. However, as we will discuss later, it is not about turfing them into a field and letting them get on with it. As we shall see, you need to get to know your animals. And they need to get to know you.

When we first got sheep, we went on a sheep-handling course. It was very interesting, part practical and part theoretical. The sheep handling was based around what might best be described as standard farming practices. Simply put, this means rounding the sheep up and moving them into a penned-off area. This is normally shaped like a wedge with a large opening for the sheep to go in. At the narrow end, there would be sheep race. At the end of the race is a bidirectional gate enabling you to split the sheep into two flocks. This is useful for separating tups from ewes, for example. The race is also useful when vaccinating or drenching the flock.

Well, if you have never tried rounding up sheep before, then you are in for an interesting experience. Animals are masters of reading body language, and the moment they know where you want them to go, they choose to go elsewhere. And they are a lot stronger and faster than they look.

One night, our race was stolen. This turned out to be a blessing. We switched from rounding up sheep to leading sheep. In essence, we started to apply animal psychology. It took a while and much patience from Nicole, our principle trainer, but over time we gained the trust of the sheep. These days most of them recognise their names and can be called into the treatment pen individually. That has surprised a couple of vets, let me tell you. This taught us a lot about animals. It is far easier to handle them if they trust you.

Furthermore, we learnt that we no longer had to "tip" sheep in order to restrain them. If a sheep is relaxed in your company, it is easy to slip on a head collar and tie the sheep up in a corner of the pen with a lead rope. We find it easier to trim bottoms with the sheep in an upright position. Best of all, we can take our time doing these jobs as the sheep are not struggling to get away and our backs aren't complaining as much either.

When we got cows, we applied the same philosophy. One cow came with a calf, so they both started out quite wary. Every day, we went out and combed the cows. With cows, you start combing at the back and work your way slowly forward. The cow will soon tell you when you are far enough forward. Over time, we managed to get to the point where we could scratch the top of their heads. The calf, over time, became curious and came to us. It was not long before he was coming in for all-over body scratches.

So, in our experience, taking the time to gain animals' trust makes them much easier to handle. A cow will only go into a trailer if it wants to. The same goes for pigs. There's no amount of pushing or shouting that will make a pig or cow go somewhere it does not want to go. But if you can lead them, they will follow.

And there is something intensely satisfying about calling the vet and watching the dismay on their face (when they see the cows out grazing) turn to astonishment as you call the cows into the crush one at a time.

This takes time. It never stops. You have to spend time

with your animals every day. If you leave them alone for long, they will soon return to feral ways. We check our animals three times a day and spend on average about an hour a day with each type of animal.

For us, this is also what it is all about. We spend time with our animals and we love it. We sit with them and they sit with us. They come over and rest their heads on our legs. The relationship you build is the most rewarding part of keeping animals.

However, getting to know your animals and building trust takes time – and by time I mean not just a few months, but years. Animals will tell you when something is wrong. But the signs can be very hard to spot. Our calf was eating haylage, roaming and seemed generally fine. Then he collapsed one night, and we were unable to save him. We didn't know cows well enough at that point in time.

Ymogen, one of our hogs, is still alive because Nicole spotted there was a problem. Ymogen was just slightly away from the other sheep with her head drooping a little. Not unusual as it was winter and the weather was cold and wet. However, there was something very subtly off. It turned out she had broken her jaw.

Yinn, another hog, was sitting in the sun with all the other lambs, but there was something about the way she was sitting that didn't seem quite right. It turned out that she had coccidiosis but was saved by emergency medication from the vet.

The smallholding life is full of such experiences. Knowing your animals can save them when they are in trouble. Knowing you did everything possible can make it easier to cope if they pass.

Of course, when they first arrive, you have zero relationship with them. It is an interesting aspect of human psychology that we make a lovely home for animals and then expect them to be wowed when we put them in it. No! You can take an animal from a dingy hovel into a luxurious setting, but it won't thank you. As far as it is concerned, you have taken it from its home and planted it in a strange place surrounded by strange humans (probably making silly noises). At this point, the animals are close to panic.

When we first got pigs, that was pretty much what we thought. We had prepared a pig run comprising a large area with a spotless pig arc full of clean straw. The piglets came from a shed. What a great new place to live, we thought. They'll love it. But they were having none of it. Within an hour, they had legged it. Luckily, we managed to catch them.

Animals need time to settle. What we can do, as humans, is respect that. We still start the trust-building exercises from day one, but at this stage it might just involve sitting there reading a book. It certainly doesn't involve approaching them.

Animals are curious by nature. If you sit there long enough and nothing bad happens, they will eventually come and check you out. Each time they check you out, it will be a little bit more intense. Of course, you can expedite the process with treats. Beware, however, as treats can also cause problems.

Animals are creatures of habit. If you bring treats daily, they will come to expect them daily at the same time every day. It is worth noting that they can become bossy and anxious if you get into the habit of always bringing them treats. Also, too many treats, such as sheep nuts or cow nuts, can cause problems in the rumen.

But animals do not understand what, to them, is an inconsistent approach. Imagine, if you can, trying to explain to a vexed Highland cow that you have no treats today.

Continuing on the theme of animal psychology, it is worth thinking about where these animals originate from – in other words, how they have evolved. While the breeds may be hundreds or thousands of years old, they are still essentially the same animals as their wild ancestors. It is a bit like dogs and wolves. When you strip away the breed characteristics of a dog, you have a wolf.

Basically, most animals like to roam. Buffalo wander on the plains. Boar wander through the forest. No amount of breeding really prepares an animal for life in an enclosure.

We have found that our animals like to make their own decision as to where to go on any particular day. They even seem to convene meetings before heading off. It is the same instinct in dogs. We humans recognise this need in dogs, which is why we take them for walks. Recognising it in farm animals can make for much happier animals.

This can be quite important as animals not only like to roam, but they also like to explore. They have not read the parable "The grass is greener". To them, the grass is greener the other side of the fence. If you provide enough space, they will be content. If they get bored, or hungry, they will explore.

Fencing should best be seen as an invitation to remain. We have witnessed a pregnant ewe climb over four-foot high sheep hurdles. Standard stock fencing is around three feet high. We have often found cows outside their designated field. Sometimes, they leave no indication of how they got there. I have never forgotten the sight of Bluebell, our Highland cow, standing outside the gate waiting to be let in. One of our tups took a gate off its hinges to get into the neighbouring field. There are many such stories, all quite funny in a way, until it happens to you.

Moving on, I have covered the need for shelters and water elsewhere. It is just worth adding that the different seasons bring different workloads. During the winter, animals will need to be fed daily. This can add to the time needed to look after them. Shade is important in the summer for all animals, even here in Scotland. Somewhere to escape from the mud is equally important in winter. A winter shelter is even better, even for the hardy breeds.

Aside from that, there are legalities to be aware of. All farm animals, except chickens, need to be registered with the respective agency. Chickens may need to be registered depending on the size of the flock. Prepare yourself for hours of administration. Each animal has a different agency, and sometimes there is more than one agency per animal. Each has different movement procedures, tagging requirements and different requirements for medical histories. Each country has different agencies. It is a pain, but one you have to go through.

Rural crime is also something to think about. Sheep and cattle rustling sounds like a crime from the Middle Ages. It is not. It is around today and in some areas it is rife. Where we lived in Somerset, for a few weeks of each year, sheep were stolen every night. Usually whole flocks are taken. Sometimes, some are butchered where they stand and the meat taken away. Sadly, it is a fact of life, but you can take steps to reduce the risk. My favourite is large prickly hedges. You can also reverse the hinges on gates making it harder to remove them. We had locks with audible alarms on some of our gates.

17

VERA, VI AND URSI

We were moving house in the spring and the planned move was only a few weeks away. In the autumn, we had thought long and hard about whether it would be safe to lamb. At that point, we hadn't found a house to move to let alone sold ours, so it was all up in the air. In the end, we decided it was safest not to.

But now we had all the dates sorted we were wondering what to do. We knew we were going to be busy and that moving house can be pretty stressful. Also, we were not entirely sure what the lambing facilities would be like at the new house. On the other hand, we would both be around (neither of us would be going out to work during the first few months of the move). This thought, and a moment of spontaneity, made us decide to buy some ewes in lamb. So we ordered three from a farm in Yorkshire. Having lambed ten the year before, we expected three would be a lot easier. The collection date was set for a week after we moved in. No point hanging about, we thought.

So, seven days after we moved, we were back on the road to collect our new sheep. We were offered a choice and chose one with triplets, one with twins and one with a single due. We loaded them into the trailer and drove back through heavy rain and high winds. We named our new sheep Ursi, Vera and Vi.

We collected our new sheep five weeks before the due date we had been given. Annoyingly, we didn't have individual dates for each sheep. This would mean regular four-hourly checks every day and night. Having lost some lambs the previous year, we were keen not to miss any births this time around. We had the lambing shed all prepared, decked out with straw and plenty of hay, water and a mineral lick. Although early, we thought it would be best to put them straight into the lambing shed. We had put our own sheep in early the year before due to inclement weather and they had been perfectly happy – ecstatic to get out of the rain, as it happens.

We were also hoping that by keeping them close by, we'd be able to spend more time with them. The better we got to know them and the better they got to know us, the better lambing was likely to go.

They seemed to settle in well. They were a bit disconcerted by the change in scenery, and also by us and our dogs. Vera, particularly, was unimpressed by our dogs. It should be said that our dogs are good with sheep. George looks to protect them. Whereas Haribo, the collie, was if anything a bit scared of them. Nevertheless, all seemed fine.

However, things soon began to go wrong. They started acting unwell, but the symptoms were not clear. It looked like it could be twin lamb disease, but this puzzled us as they had plenty of hay and were also on a carefully measured supplement of high-protein sheep nuts. Twin lamb disease would indicate that they were not getting enough to eat, or not enough nutrition. That said, Ursi was expecting triplets and that, combined with moving to a new farm, would have caused some stress to all three ewes, so we didn't rule out the possibility.

We called out the vet. She could find nothing wrong and said it might be worth treating them for twin lamb disease anyway. We could tell she wasn't convinced it would help, but we tried. The vet turned out to be right; it made little difference. We were stumped.

We were also a bit under the weather ourselves. With the move, the purchase of pigs as well and the constant checks on the sheep, we were run ragged. Then things got worse. Vi started to go downhill. We called the vet who thought Vi might be suffering from calcium deficiency, so she was given a calcium injection. This is a big injection: 50ml of calcium solution, which the vet split into two of 25ml each and injected subcutaneously along Vi's ribcage on each side. Vi perked up slightly, but the next day she had gone downhill again.

The three girls were clearly in trouble, and we had to do something. The problem was, what to do? After some discussion, Nicole remembered that last year she had taken some of the sheep for a walk so they could access better grass than was available in the paddock (which by now was quite churned up).

Nicole had been halter training the sheep, so she haltered Ursi and took her for a walk. Ursi went straight for some grass. We took that as a sign. We had a small paddock we could put them in. It had shelters so we could put down some straw for them. It was also small so we could contain them for the checks. We took them over and let them into the paddock. All three went straight for the grass and were tucking in big time. In fact, by the next day, they all looked a lot better.

We thought about this and came to the conclusion that they had not been eating enough hay. The hay had been left by the previous owners, and we had assumed it was fresh. However, we think, in retrospect, it might have been a bit too old. That would have meant that the problem would have been acidosis rather than twin lamb disease. Same symptoms, opposite causes. If they were not eating enough hay, they would be eating too many sheep nuts, and this would affect their rumen. That was a lesson learnt.

We were, however, not out of the woods. Suddenly, Ursi went rapidly downhill. She was showing different symptoms from the last time, though. One minute she had been grazing away happily, the next she had collapsed. We knew there was no time to waste. We would have act fast, and there would be no time to wait for the vet. I stayed with Ursi while Nicole ran back to the house to fetch the big syringe and bottle of Calciject. This would be the first time administering such a big injection, but Nicole was strangely calm as she administered the two injections, one on each side of Ursi's ribcage. They seemed to take forever, as 25ml is a lot of liquid. It is funny how time distorts in these situations. But the results were miraculous – Ursi was back on her feet in a matter of seconds as though nothing had happened. We breathed a sigh of relief and retired for a well-deserved cuppa.

No sooner had we sat down, or so it seemed, than Ursi started to go into labour. It was a week early, which was a worry as premature lambs are hard to keep alive. We brought all three back into the lambing shed and monitored Ursi. Nothing was happening, so we called the vet. He confirmed that she was in labour and set about helping Ursi to give birth. He delivered three lambs, all tiny and unable to stand, let alone suckle. We named them Bert, Pinkie and Blue, the latter after their woolly jumpers we'd put on to keep them warm. I took the smallest, Bert, inside and put him in a basket in front of the fire. Nicole tried to get the others to latch on but was only having success with one.

We had to resort to the tube feeder. This is not something we like to do, but we followed the directions in the books with regards to dosage and time between feeds. I stayed up all night with Bert, sending Nicole off to get some sleep. Sadly, he died the next morning in my arms.

After only a couple of hours sleep, Nicole resumed helping Blue and Pinkie to latch. Pinkie was the stronger and, bit by bit, she got there. However, Blue was just too wobbly, so we had to continue with tube feeds. Sadly, she also passed, this time in Nicole's arms. In retrospect, we think she may have caught watery mouth but we missed the signs. When things like this happen, you spend a lot of time wondering "what if" and beating yourself up for the things you did that were maybe wrong and the things you should have done but didn't. Blue probably did not get enough colostrum and was therefore lacking antibodies.

Truth be told, this was not going well at all: three lambs born and only one surviving. But we still had two pregnant ewes to care for, so we had to carry on.

It was not long before Vi looked like she was going into labour. But she, too, was showing signs of having problems, so we had to call the vet again. We were more than a bit worried as Vi was giving off a strange smell. The vet confirmed our worst fears: the lamb had died in utero. This was devastating not only for us but for Vi. Of all the three, she had shown the strongest maternal instincts gazing longingly at Pinkie. Poor Vi had to have an emergency caesarean then and there. Nicole held her head, and I held her rear end to comfort her while the vet did her magic.

On top of this, we had to separate Vi from her friends and put her in with the main flock. As animals often do, they did not welcome her with open arms (or legs). Even sheep can be a little cruel, so poor Vi was on the receiving end of a bit of bullying. There was nothing we could do, but it added to the stress we were feeling.

We also had a further problem – what to do with Vera. Ideally, we wanted to keep her in the lambing shed to give birth to her lambs. But she was on her own now. Ursi was out with Pinkie in the small paddock. You can't keep the newborns in the lambing shed for very long, as it is not hygienic enough no matter how clean you keep it. So, Vera was alone in the lambing shed, and she was not happy. She was making a lot of noise, so I went over to see if I could calm her down a bit. By now, with everything we were trying to get done, I was also not in the best of moods by that point.

I went to Vera, but she was just not interested; and then, in front of me, heavily pregnant, she somehow climbed over the sheep hurdles and looked ready to run for it. Amazingly, we had two strokes of luck. The first is that Vera, having now escaped, wasn't sure what to do next and just stood there looking mildly surprised. The other is that Nicole happened to be on her way to the shed with a halter and lead rope. As quick as a flash she popped the halter onto Vera, and we led her along the track, through the big hayfield and down into the paddock where Ursi was with Pinkie.

This brought more worries for us as there were crows nesting near the wee paddock. When Vera decided to give birth, the last thing we needed was the crows getting there before us. So, we upped the checking schedule. It was freezing cold and the paddock was about half a mile from the house. Those night-time walks through the frosty air are etched into both our memories. It was exhausting, but we had no choice but to keep going.

Vera was also taking her time. We wondered if there might be some other sign we could look for that might tell us she was imminent. Nicole scoured the internet and discovered that ewes can subtly change shape around 24 hours before giving birth. It is a subtle sign, a change caused by the lambs moving from the flanks. In effect, the sides seem to drop. At this point, Vera's were definitely bulging out. We became very familiar with Vera's flanks during our checks.

The May Bank Holiday came and along with it some welcome sunshine. It was still cold, but bright sunshine does lift the spirits.

Nicole came back from the early morning check with news that Vera's sides appeared to have dropped. It was, as had been described, quite a subtle change, but noticeable for all that. For us, this was great news; Vera could be giving birth within the next 24 hours.

Of course, life is never quite that simple. Nicole had a new gardening client that afternoon. Luckily, it was not far away, so Nicole could get back quickly, if needed. I decided to stay in the paddock with Vera while Nicole was working. The ever-present threat of crows was on my mind. I had a good book with me and settled down next to the stone dyke.

Looking up from my book for the umpteenth time, I noticed that Vera's behaviour had changed. Again, it was subtle. But with all the checks, we were getting to know Vera quite well by now. I put my book down and went over to take a look. Vera still wasn't that sure of me, so I couldn't get that close, but I managed to get close enough to see a spot of fluid. The waters looked like they might be breaking. I called Nicole, and she set off home straight away.

Almost as soon as Nicole arrived, Vera went into labour. With no help from us, Vera gave birth to her first lamb. This is how it should be, we thought. The sun was shining and a healthy wee lamb had been born. We helped Vera dry her off (she was a girl). Within half an hour, Vera gave birth to a second wee girl. Vera truly was a master at this. Nicole helped the lambs find the teat, and we sat back to watch.

Finally, lambing was over, but we were exhausted. Rather than being easier, lambing three had proved more tiring than lambing ten the previous year. But we had learnt a lot. We vowed that next year was going to be different. Vi had lost her only lamb, Ursi had lost two, and last year Sparkle and Peaches had lost both of theirs while Scarlett had had such a terrible birth that she'd abandoned her two. We felt that we owed these sheep.

As soon as we thought the three lambs were up to it (sometimes there is a bit of argy-bargy from the non-mothers), we brought Vi back from the main flock to be with her friends again. Well, Nicole did. Vi was so ecstatic that she did the most amazing jumps Nicole had ever seen a sheep do. This confirmed what we had always thought: sheep form strong bonds amongst one another.

As the lambs grew, we merged the two flocks into one, but, for many months, the Yorkshire girls tended to stick together. Over time, though, they became fully integrated and are now one happy flock.

We put what we had learnt into practice and built new lambing facilities with a large shed, plenty of grass and electricity. The following year we had a much better lambing, and all three Yorkshire girls had successful births and produced lambs that grew up healthy and strong.

18

LAND MANAGEMENT

One of the biggest challenges facing agriculture in the twenty-first century is land management. Ever intensifying production and the trend towards monocultures are putting intense pressure on the soil. This affects all of us – large farms and smallholdings alike.

As a smallholder, the type of land which comprises your holding is a key part of your decision as to how you can utilise it effectively. In Britain, we have a huge range of land types, from salty and windswept coastal areas, lush meadows and marshland through to mountains. There are the flat plains of land reclaimed from the sea in places such as Somerset and Norfolk through to the rugged hills of Yorkshire and Lancashire. The west of Britain, especially Wales and Western Scotland, have higher rainfalls than the east. The land itself can be chalky, peaty, stony, and so on. All of these areas require different approaches.

Historically (up to the mid-twentieth century) much of farming was small farm based. In Scotland, we had the crofts. Local communities worked together to raise animals, grow crops, cut hay and produce their own food. Surplus food was sold in local markets to local people. People were close to the land and managed it based on knowledge gained over centuries.

This small-scale farming was based around a cycle of changing land use on a regular basis. A field would have crops one year, sheep another, vegetables another year and perhaps cows after that. The key thing is that there was constant variation. What was taken out of the soil by crops was put back by the animals. The soil had a chance to recover due to it not being ploughed every year. In essence, what happens at the surface of the soil has a huge impact on what goes on beneath.

Soil is a hugely complex microsystem, but, for the sake of this discussion, let's simplify things and describe it as a mixture of rotting organic material and sand. It is managed by worms. If the worms go, the soil will become compacted and cease to be aerated. In this form, few plants can grow.

Led by farm subsidies and the ever-increasing pressure from supermarkets to drive down prices, farming has evolved into more of a business than a way of life. At the most damaging end of this is the "factory farm".

Intensive factory farming is bad news for the soil. Growing barley and wheat year after year in the same field leaches the soil bare. Pouring on fertilisers might boost nutrients a little but does nothing to boost organic matter. These nitrates also leach into rivers causing all sorts of secondary problems. Constant ploughing leads to a breakdown of the soil structure and erosion, the soil being blown and washed away.

Intensive livestock farming means huge amounts of animal waste being concentrated in small areas. In essence, the soil is turned into slurry. Too many animals in too small a space compacts the soil, essentially rendering it into something that resembles a concrete block.

As a smallholder, it is important to think carefully about how you can best manage your land.

The topography is important. It is nigh on impossible to grow crops in hilly fields. In some places, slopes have been converted into terraces – tiered, horizontal areas in which crops can be grown. But this is not always practical. Many hilly areas such as those in the Scottish Highlands and

Lowland Hills, the Welsh Hills and the Lake District can only support permanent grass pasture or woodland.

This is also true of coastal locations, such as the Scottish Islands and coastal areas of Britain. It would be difficult, if not impossible, to grow crops in many of these areas due to the high levels of salt and the constant battering of the elements.

So if, for example, you have a holding with permanent grass pasture, then it is likely to be grass pasture for a reason. Switching it to growing crops could be courting catastrophe. For example, our holding is in the Scottish hills, and if you were to plough up one of our fields, the winter rains would wash all of the soil away by spring, and you would be left with bare rock.

As a smallholder, you can make a difference. There are a number of methods out there for managing land, but in essence our approach is to look into how best to implement traditional methods in the modern day whilst bearing in mind the type of land we have and what is best suited to it.

If you have permanent grass pasture, you need to manage the grass. The traditional way to do this is with livestock. Grazing different types of animals at different times is a much more natural approach. Cows may get a bad press for their methane production, but in the right places they are an integral part of managing the land.

Further, if you are looking to grow food or make money from your holding, you need to convert the grass into something edible for humans. The only method I am aware of currently that can do this is to graze ruminants who basically eat grass and turn it into meat.

If livestock is not your thing and you are looking to grow crops, this needs to be factored into your choice of holding. You will need somewhere suitable for growing crops.

Traditional methods are based on crop rotation combined with periods where the land lies fallow. This approach enables you to replenish the soil with organic matter and nutrients. This, in turn, feeds the worms and keeps the land fertile and productive.

In summary, all of this information needs to be taken into account when you decide what you want to do with your holding. And it can be quite complex. It is a balancing act.

Modern farming relies more and more on extracting more food from less land. It is the holy grail of higher productivity. With pasture, this tends to involve the use of weedkillers to kill everything but the grass and fertilisers to boost the grass growth. Even with this approach, keeping too many animals can lead to overgrazing, which, in turn, can be catastrophic for the land. In the Outer Hebrides, for example, overgrazing in the past has led to a breakdown of the root structure which has resulted in coastal erosion.

The same applies to crops. If you have experience of growing vegetables, you will know that they are spaced out for a reason. Planting vegetables too close together means they just don't grow very well. Also, the soil needs to be fed with both organic matter (compost) and fertiliser. The latter can be achieved by planting nitrogen-fixing crops combined with a general fertiliser.

So, managing land without chemicals and avoiding overuse means less animals or less intensive planting. This means less food and less income. It is all part of the balancing of time, money, effort and resources that are fundamental to smallholder life. It is something to think about.

Here in south-west Scotland, we can only grow vegetables, keep animals or grow trees. Making a living from growing trees is something of a long-term activity, as they take a while to grow and they need a lot of land. It is not really an option considered in this book. That said, half of our land has been converted into natural woodland. Our intention is to coppice this to provide firewood as well as areas rich in wildlife. Whether we have enough firewood left over to sell remains to be seen.

We make an effort to look after our pasture. As it is permanent grass pasture, we only use it to graze livestock. Our focus is sheep who like to nibble the grass down to its roots. To manage our patch, we keep the numbers of sheep down to avoid overgrazing and we borrow cows from a neighbouring farm from time to time. This has two key benefits: The first is that by keeping the animal population low, we also keep the worm egg count down. Also, by mixing sheep and cows, we disrupt the worm cycle for both; thus, benefitting both.

Cows prefer longer grass so they open up wider areas for the sheep. Cowpats are also highly nutritious for the soil, so this means we don't have to use chemical fertilisers to keep the grass in good shape.

This is a highly complex subject, and there are many theories and ideas in the public domain. It is worth carrying out a bit of research and factoring land management into your plans.

19

UNINVITED BADGER

One of the first things we did on moving to south-west Scotland was to move the chicken coops up closer to the house. As well as making it easier to look after them, it means we are also close by if there is trouble.

One night, Nicole was woken by sounds from the chicken coops. They were squawking and generally making a terrible racket. Something wasn't right. The hen coop doors were closed – we have electric door controls that open the doors at dawn ad close them at dusk. While this renders them fox proof, our first thought was still "fox".

We threw on some clothes and went down to investigate. We brought George, our large dog, along in case it was a fox. By now the noise had stopped, so we were a little puzzled. We opened the nest boxes to take a look. All seemed quiet, just the usual bundle of hens blinking in the torchlight, but nothing untoward. We closed up the nest boxes and went back inside.

No sooner had we settled down than it all started up again. Back we rushed, George in tow, for a second look. This time we took the roof off. Having a plastic chicken coop has its uses; the roof can be removed quite easily.

We peered in, but once again all seemed normal. The hens were just perched there blinking in the light as though we had just woken them. As we looked in, I noticed movement. All our hens are grey in colour, but a grey hen-sized bundle on the floor of the run was slowly changing shape. It was a badger uncurling itself. I think, in surprise, a profanity might have escaped my lips.

It was quite a small badger so probably quite young. It looked extremely cute and cuddly. Indeed, it looked so cute that Nicole went to pick it up. I quickly intervened, probably shouting something like "No". Badgers might look cute, but they are formidably strong. The last thing you want is one biting you. Cute or not, it was still a wild animal and most likely felt threatened by us. We also hurriedly got George out of the hen run. George might be big and powerful, but even a small badger like this could inflict a serious injury.

By now, the badger was moving around the henhouse looking for a way out. It was also in a bit of a panic – we had given it something of a fright. In turn, the hens were now panicking and flying in all directions. We opened the henhouse door to let the badger escape. However, it was the cockerel who was first out through the door followed by three hens. At the back of my mind I was thinking that surely the cockerel was there to defend them, not run away.

The badger, at this point, had climbed up and over the sides of the chicken coop and escaped through the gap where the roof used to be. It ran down the hen run along the stone dyke. George, suddenly aware of what was going on, gave chase – on the other side of the wall, thankfully. Both disappeared into the dark. We called George back just in case.

With all that excitement over, we took a closer look. Sadly, there were two dead hens. One was Mrs Mills, the mother of 13 chicks just a few months before. Another was Petal. As the youngest, she had been bullied quite a lot in the past but had grown in confidence here with all the space and other hens in the flock. It was very sad. We placed them gently in a box and put them carefully to one side.

Hattie, Mrs Mills' mother, was injured so we took her indoors. We also found Jane trapped under one of the roosting bars. We took the bar out freeing her and took her inside as well, just in case. We closed up the henhouse and went in search of the runners. We couldn't leave them out, not with a badger about.

Well, if you have ever tried to round up frightened hens in the middle of the night, you would know the daunting task that lay ahead of us. We did manage to catch one who had somehow contrived to *corner herself. But the other three, they just kept running off in different directions into the dark. The search was not helped by the proliferation of brambles. There was also the concern that the hen run was only surrounded by a one-metre high stock fence. A spooked hen could easily fly over that and be long gone.*

It took about two and a half hours, but little by little we managed to coax them back into the chicken coops. Of course, they were wary of going into the main chicken coop they had just escaped from, but fortunately we had two others. We opened the doors and managed to coax them in one by one. Finally, they were all back inside with the doors closed. As a final measure, we tied the door shut and leant a heavy pallet against it. They were as safe as could be, badgers aside.

We managed to catch some sleep but were up early to check on things. No more attacks, thankfully. It turned out that Hattie had quite a bad injury: she had been bitten under the wing and needed stitches. We took her to the vet and left her with them. In the meantime, we went shopping for anti-badger measures and ended up buying a mains-powered electric fence. More expensive but better than batteries. We also bought a voltage checker so we could check the fence was working.

Sadly, Hattie never recovered from her operation. Hens and general anaesthetics don't go well together. We collected her so we could give her a decent burial.

In the cold light of day, we pondered on what had happened. As it turned out, just a few weeks before, our neighbour had lost all but one of their hens to a badger. They had set up a camera and caught him coming back nightly. He was a large dog badger, not the young badger that had been in our run. So, there were two badgers on the loose. We were in the middle of a prolonged cold spell and there was snow everywhere. We suspected that the badgers were roaming further in search of food. You can't blame them really.

We also set up a wildlife camera, but we never saw the badger again. Perhaps George had done his job and scared it off.

It remains a puzzle to this day as to how the badger got into the chicken coop. There was not a scratch anywhere. Given that badgers tend to dig their way into things, I would have expected to find claw marks. The only logical explanation is that the badger had somehow opened the door. Perhaps it had managed to get a claw in between the door and door frame, although this would have been tricky as the door fits snuggly inside its frame to prevent exactly this from happening. We'll never know.

The mains electric fence has now been wired in by an electrician, and we have it on a timer. Wouldn't do to have the henhouse doors open while the fence was still on.

It also highlighted that Nicole had been right to bring them closer to the house. Had the henhouse still been 200 yards away, in its original position, the chances are all the hens would have died that night.

Nature can be cruel.

20

WHEN FRIENDS AND RELATIVES COME VISITING

Many city dwellers enjoy weekend breaks in the country. Having lived in a city for much of my life, it was quite common to find me heading out to the country for breaks. As a child, I was often out camping in the Scottish countryside. Family holidays were mostly in the country or by the sea. As an adult, I enjoyed exploring rural areas across the world.

So, when you move to the country, many of your city friends will be jumping for joy as they now have a new holiday destination. In a way, it is a free break as there are no bed and breakfast or hotel bills to pay.

In fact, we had barely started unpacking when the first of our friends invited themselves to stay.

To be honest, we do like having people to stay. As well as breaking up the routine a bit, it means we can show off what we are up to. We are proud of what we have achieved, so we like others to see it. However, there are good points and bad points when friends and relatives come to visit.

A somewhat perplexing challenge is that many have a somewhat idealised view of smallholder living. We often get the feeling that people think we don't have much to do here. *What is there to do?* they seem to ask themselves. *The animals look after themselves, a bit of gardening, what else?* We get the impression that they think we have a lot of free time.

The reality is that with both of us working, if you add in the hours spent on the smallholding, which is essentially a job, we are both working longer hours than when we were in corporate life.

Smallholder life is hard work. That is part of what this book is all about. The fact is that animals do not look after themselves, nor does the garden or the vegetable patch.

One of the first thing that surprises our guests on their first visit is when Nicole disappears off to work. Nicole works as a gardener during the week. It feels like people assume we gave up work when we took on this life. Nothing could be further from the truth. I also work but some of my work is computer-related. That means I can be more flexible and therefore more available to be with our guests. What they don't see is me madly catching up once they have gone.

If you are going to stay with friends that have "normal" jobs, you expect them to be out all day if you are there during the week. People who come here expect us to be in all day. It can be a bit frustrating.

This kind of attitude is also prevalent with "drop-in" visitors. These are people who drive over to see you without telling you they are coming. They just expect you to be there and to have time on your hands. They can arrive, see you outside working flat out, covered in mud or dirt or paint or something like that, and then proceed to take up the rest of your day while you run around making them cups of tea.

Others come to treat our smallholding as a sort of petting zoo. Much as we take pleasure in introducing friends and family to our sheep, we are very aware that sheep are by nature flight animals, so not very comfortable around people they don't know and sometimes they can get a bit stressed. It is hard to explain this to an enthusiastic five-year-old without sounding like an old stick in the mud. But, at the end of day, the welfare of our animals always come first.

Sometimes complete strangers even ask us to take them to see the animals. We can still remember someone picking up sofas we'd listed on Freecycle expecting to take up most of our day looking around and meeting our animals. Telling them that you are busy or that sheep can be a bit spooked by people they don't know makes little difference – "oh, they won't mind me" is often heard.

As well as working, you have many farming tasks to get done. Animals can get themselves into all sorts of pickles. It may look as though they spend their days nibbling at grass and generally doing nothing, but that's not the case. Farm animals can be quite high maintenance. Like us humans, they can get sick from time to time. Unlike us humans, they can't speak, so you have to observe their body language to ascertain when something is wrong. In order to do that, you have to get to know all the animals. This all takes time.

When things go wrong, it can eat up huge amounts of time and send any plan or timetable out the window. It is all part of the fun.

On top of this, there are all the routine repairs to infrastructure, not to mention construction of new infrastructure. In the last two years, we have rebuilt walls, built two field shelters, two sheds, put in miles of fencing and gates, and pulled out tons of brambles. It is a long list.

The point is, nobody really sees this and when people come to visit or stay, most of these activities are put on hold. One of us will be out doing the animal checks while the other is entertaining the guests. We have found that we are happy to do this for a short period, maybe two to three days. After that, we have to get on with what we do, which takes most of our time every day. At that point, guests become an additional activity; one we don't really have the time or energy for. It is also a pain in the neck planning menus for more than a couple of days. The thing is, if you live "the good life", you feel you can't just roll out any old food. There's pressure to put out a proper farmhouse spread – home-made this and that and big quantities plus an array of homebrew. All this is lots of extra work.

Another thing to watch out for is the fussy eater. Having put in a lot of effort producing a farmhouse spread showcasing your carefully prepared home-grown produce, it can be exasperating to have someone pick through it as though it were monkey eyeballs.

That said, on the plus side, sometimes people come to stay with the intention of helping. My brother-in-law came for a week once and helped to build our lambing shed. That was brilliant. If this were the norm, we'd be tempted to have guests full-time. To be fair, my brother-in-law was glad of the physical work as he'd just split from his girlfriend and was glad of the distraction!

The key here is to be ready for this. People coming to visit or stay can be great fun. It just needs to be properly managed. We are a bit stricter now about how long people can stay. We also encourage guests to go out exploring on their own during the day. This takes the pressure off us a lot. We now feel much more in control and, fussy eaters aside, we really enjoy having people to stay.

21

BORROWING COWS

Being a hill farm, all of our agricultural land is permanent grass pasture. We keep sheep and they have plenty of grass. However, even grass pasture needs to be managed (see Land Management); otherwise, it will degrade over time. One of the best ways to do this is to graze different ruminants.

With this in mind, we did, for a while, own three Highland cows. However, we soon found that we just did not have the time to manage so many different types of animals. Sadly, we decided to sell our cows.

Cows and sheep go really well together for a number of reasons: First, cows eat the long grass, whereas sheep like to nibble at the shorter grass. With cows grazing the grass, they open up areas that would otherwise be ignored by the sheep. Cows and sheep are infected by different worms, so if you graze cows on the grass followed by sheep, it naturally cuts down the parasitic worm count. Finally, cow dung is also an excellent fertiliser for the grass, so cows act as a natural fertilising agent.

This year, the weather was perfect for grass. It just grew and grew and grew. There was so much of it the sheep could not keep up. Our neighbouring farm keeps cows, and one day we saw them in grazing in a paddock bordering ours. We thought it would be a good idea to ask if we could borrow them for a few days. It was a win-win for all of us: the cows get new grazing, the sheep benefit, our neighbour's fields get a bit of a rest and we get the benefits of cows without having to own them.

Our neighbour readily agreed, and we set a date for later in the week, a Thursday evening as it happened, for them to move in. The cows were set to be on our fields for around five days. Perfect, we thought.

The Thursday evening came and our neighbour moved the cows into our field. Like all animals in new places, they set off to explore. We had put them in our largest field which had a large steel water trough. It was a good-sized herd with a mix of cows, calves and a large bull – around 30 in total. After a good look around, they soon settled and all looked well.

The next day I was scheduled to spend some time over in Edinburgh with friends and would be staying over Friday night. It was a good trip until Saturday morning when I received messages from Nicole telling me that the cows had run out of water. In fact, when I got home a couple of hours later, we had a full-blown cow water emergency on our hands.

Nicole had been taking water over but with limited success. While we have a 600-litre bowser mounted on a trailer, it can take forever to fill up using the tap. Nicole had been ferrying over small amounts and trying to get them into the water troughs and supply tank, but it was proving to be too little and too slow.

The weather had been hot and sunny for quite a while, so things were drying up. We have two water tanks buried in our fields that collect spring water and deliver them to water troughs dotted around the fields. We had been having ongoing problems with a leak in our large field tank and although I had taken advantage of the dry weather to repair this, the tank had remained empty because the stream had dried up in the hot weather. The lower tank, smaller at 400 litres, was now empty, too, and its supply stream was also dry. On top of that, the cows had drunk dry the water troughs to which they had access.

We are on a private water supply here. A single tank catches water from a stream and supplies ours plus three other neighbouring houses. I checked our water tank; it was only three-quarters full. We could not use this.

We have a communal water pump for water emergencies, which had been seconded by one of our neighbours for watering his polytunnel. I set out to look for the pump, as we'd have to pump up water from *the river rather than use the taps. That neighbour was out. I couldn't find the pump, so I asked another neighbour for help. Finally, he tracked it down and we set about connecting it all up. All the hoses and related items were sited at the neighbour's house, so we set it up there. We try to be neighbourly here, so we thought our absent neighbour wouldn't mind.*

We had to connect it to his electricity supply as our houses were too far away.

Having got it all going, I set about filling up our water tanks. Once full, the 600-litre bowser can be towed by the tractor. It takes about half an hour to fill using the pump. While I was there, I also filled two 20-litre water carriers so I could top up the water troughs. This turned out to be a good idea.

With the first 600 litres, I set off into the fields. I stopped off at the large field where the cows were and carried over the two 20-litre carriers. Some of the cows were waiting by the water trough. One brown cow, in particular, seemed quite put out by the lack of water. She watched me intently as I poured the two water carriers into the trough. No sooner had I finished, she then put her head in and drained it dry again.

OK, I thought, I need to do that again. I towed the 600-litre bowser up to the main tank, connected a hose to it and left it to fill the tank. It is an 1800-litre tank, so it would need three trips. However, with only gravity acting as a "pump", it took over two hours to transfer the water from bowser to tank.

In the meantime, I grabbed the quad bike, having left the tractor with the water bowser. Because 600 litres of water is heavy (about 600 kilograms) and the trailer has no jockey wheel (to raise the front of the trailer), the trailer had to remain attached to the tractor. With the quad bike and the two 20-litre carriers, I set off to refill them and deliver the water to the cows. When I got to the water trough, a bovine queue had formed. At the front of the queue was the bull. His head was almost as wide as the water trough. As he stood there, just the other side of a rickety fence, gazing at me with slightly perturbed eyes, I did have a momentary tremor or two.

Nevertheless, after I poured in the first 20 litres, his head was straight in. In fact, he drank while I was pouring in the second 20 litres; something I found surprising as I thought the noise and splashing might put him off. But no, he quickly drained the trough and then took another long look at me. I must admit, I could not figure out what he was thinking but didn't hang around to find out. I set off for more water.

One of the problems with these water troughs is that the inflow is regulated by a ball and cock, not much different from those found in domestic toilets. The inflow can get a bit blocked up and are difficult to clear. The inflow from the tank, which was now filling, was just a trickle. I had to keep going.

In all, I made nine or ten trips with the quad bike before all the cows had had their fill. I have to admit their orderly queuing was quite a sight to see. I can't imagine what it would have been like had they all piled in. But they seemed to sense I was on their side, so they waited patiently for me to bring water to them.

By the end of the day, the trough was full again and I had managed to get 1200 litres into the top tank. Emergency over, or so I thought.

The next morning, Sunday, was Nicole's lie-in day; it was my turn to do the morning checks. The sheep were all fine, so I headed over to check the cows. The first thing I found was that one of the cows was no longer in the large field. She was on her own in the small paddock where we have the sheep field shelter. This shelter was built with sheep in mind, so it is not that tall and the openings are quite small. We didn't really want cows in there in case they accidentally broke it.

Investigating further, I found cowpats everywhere. It looked like there had been a lot of them there during the night and all but one had sneaked back before morning. Only one of them had remained, giving the game away. Whilst looking to see how they'd got in, I found a heap of hurdles, which had been spanning a gap in the fence, trampled.

Although it was Nicole's lie-in day, I sent her a message with an update. She was straight over. After much discussion, we decided to give them access to two more fields. We opened the gate and let them through. To be honest, it worked out well as they were straight off into the marshy area of one field to find interesting plants to eat. All seemed well again.

Monday morning, we found the water troughs empty again. I checked the tanks, but they were empty, too. What was going on? There's no way they could have drank all that water. It turned out that, probably frustrated by the slow refill rate, they had managed to dislodge one of the plastic troughs from its stand, thus tilting it. The water had simply run out of that and drained the entire system. Here we go again, I thought.

By now the absent neighbour had returned. It turned out his community spirit was not what I had thought. In fact, he was hopping mad that I had used his electricity to pump water for our neighbour's cows. Turns out he has a long-standing disagreement with that farmer. Oh well, time for plan B.

As I mentioned earlier, we have a pipe system leading from the river into the domestic water tank just in case of emergencies. It was installed years ago. A repair had been made to the pipe near one of our sheds. It was a simple repair, just a sleeve. It was easy to take apart and provided me with a pipe that could reach the bowser. I fetched the pump and set it up on this pipeline. This time, I would be using my electricity.

I delivered the bowser full of water to the top tank first and 40 litres to a trough. On my second visit, I could see no water coming into the troughs from the tank. My heart sank. I started to look for the problem. I worked my way up the hill checking every join. They were all fine. Ultimately, I was left with the repair I had carried out to the large field tank. I had only just filled in the trench; the tank attachment and repair were now buried under four feet of rocks and sand.

In something of a mood, I dug it all out again. However, on revealing the pipework, there was no water to be seen. The problem was not here. I sat down to ponder...there was water in the tank, but it was not reaching the troughs. The only logical explanation was that there was a blockage somewhere.

It was time for a cup of tea. Sometimes, when there are problems like this, you can stop thinking clearly and just make things worse. I recognised the signs in me, so I walked back to the house and put the kettle on. Afterwards, we both went out to see if we could figure out what to do. And guess what, water was flowing from the tanks into the troughs. At that point, I remembered that the repair had caused the out pipe at the tank to rise up slightly – the tank needed to fill up a bit more before the water flowed. I breathed a sigh of relief.

My relief was short-lived. Just in case I had not reassembled the water system correctly, I checked all the joins again. There's nothing quite like water flowing through a system to show up any leaks.

At the first join, I noticed a huge puddle. While the pipes were underground at this point, I had, in a moment of foresight, covered the join with rocks so I could get access easily. Sure enough, it was leaking. I took it apart and reassembled it. But whatever I did, it just kept leaking. I sat back and studied it and came to the conclusion that the ground must have settled and pulled apart the two sections of pipe. I fetched a piece of pipe and a straight connector and inserted a section to make one of the pipes longer. For good measure, I also replaced the leaking connector.

Thankfully, that worked. I wasn't finished, however. Over the course of the day, I filled up the top tank with three bowser loads of water and also topped up the secondary tank. Once again the water system was full and functioning, and the cows had enough to drink.

After spending three days running after cows, I was shattered. All the benefits of a night away were long forgotten. Thankfully, the cows went home soon after.

Ironically, we do currently have under construction what we call a "river paddock". There are rivers running through either side of our property, and we already have an area fenced off. The river flows through this extension. However, it is not yet cow proof; part of it is bordered by a drystone dyke that I need to repair. Before we have cows again, we're going to get that field "cow ready" so that, in future, they'll be able to drink from the river.

On the plus side, the cows did their job and ate a vast amount of long grass. When we gave sheep access to the large field again, they were off into areas of sedge grass where they never normally go. There must have been some tasty short grass left by the cows.

We will certainly have the cows back again.

22

THE GOOD LIFE - FINDING THE BALANCE

There are so many good reasons for choosing a smallholder life. Top of the list is likely to be health – not only from the cleaner air and more exercise, but also from your ability to create your own quality food.

A smallholder life means you will be spending a lot of time outdoors. Whether it is tending your vegetables or running around after livestock, you just end up exercising more. It is a lot more fun than running on a treadmill in a gym.

Much modern food from supermarkets is laced with sugar, salt and various chemicals. Even salads can contain traces of herbicides and pesticides. Growing your own means that you know exactly what you are putting on your plate. If you are a meat eater and produce your own meat, then you know the provenance. You know just what kind of life the animal had.

It is important to choose the right holding for you. If you plan to plant crops, then a hillside smallholding with permanent grass pasture would not be any good. If you plan to keep animals, you need to work out the acreage required. You also need to research the most suitable animals for your land.

It is also important to look at the infrastructure. It may look like it has everything, but it more than likely does not.

If it is set up for sheep but you want to keep cows, you have a lot of work ahead of you. In our experience, it can take around two years to turn a smallholding into one that works for you. This includes building sheds in the right place, sorting out the paddocks, irrigation, drainage and laying tracks amongst other things.

If you are moving to your holding, you also have all the usual moving stresses on top. And if you are looking to renovate your new home in some way, you are going to be busy.

The difficulty is that it is not a cheap lifestyle. Without levels of expertise honed over years of experience, you are likely to be reliant on help from others, and this costs money. Examples might include vet visits at lambing time, paid labour to fix fencing or walls, mechanics to fix the tractor, the list goes on.

It is also incredibly hard to make much money from a smallholding. You can make money, but it is nigh on impossible to make a living.

It is a double-edged whammy!

Britain is an expensive place to live. It is pretty much impossible to find a place that does not incur council tax. On top of that, there are all the expenses of modern living: transport, heating, hot water, foods you can't produce yourself, phones, TV licence, broadband, and so on. While you can eliminate many of these costs by going further and further "off-grid", there are some expenses you just cannot avoid. There are also some things you can't live without.

On top of this, there will be all sorts of equipment you will need.

This means you have to earn money. Pretty much everyone we know living this lifestyle has some sort of job too. Many are self-employed; this is probably the best way to be as you can control your time. However, many have jobs in shops, bars, hotels, and so on.

This is fine so far as it goes, but it brings up the subject of time. The more self-sufficient you wish to be, the more time you will need to spend on your holding. Looking after animals takes time. Sheep, especially, are high maintenance, but proper animal husbandry takes time every day of every year.

Growing vegetables also takes time. It is surprising how fast weeds can colonise your pristine vegetable plot. But the battle with pests is ongoing. If you prefer not to use chemicals, then it takes time. One of the forgotten aspects of time and vegetables is the time it takes to harvest and clean them. We have all got used to buying vegetables that are ready to cook and eat. They take seconds to chop up and prepare. Your home-grown variety comes out of the ground dirty, damaged and needs to be inspected for all sorts of things including slugs and caterpillars. It takes time; something you might not have after a long day.

Then there is the infrastructure and maintenance that needs to be carried out. Painting sheds, mending fences, building new structures and paddocks, and installing water systems. There is always plenty of things to do.

It doesn't take much to turn running a smallholding into a full-time job.

The hard part is balancing all this. The more you work in a job to make the money you need, the less time you have on your holding – and vice versa, the more you take on to do on your holding, the harder it is to fit in fee-paying work. If you are not careful, you can run yourself ragged and get pretty stressed out.

This is particularly true if you choose to breed animals. Lambing, for example, can be a period of almost full-time work for anything from four to six weeks. Caring for sick animals can eat up days at a time. As a smallholder, you will most likely find yourself getting quite close to your animals and, therefore, find yourself devoting quite a lot of time to them.

A bit of planning goes a long way. This means both

financial planning as well as planning for what you are going to do with your smallholding. Some things have pretty fixed dates. For example, lambs are usually available late summer and early autumn, while bees are only available late spring and summer. Other things are more flexible. If you can plan your way into this life, you can keep the stress levels down.

So, to summarise, this smallholding life does have its own stresses. To be honest, it is a lot less stressful if you don't plan to keep livestock. Crops, as a rule, don't need to be tended every day, but animals do. In the winter, they need to be fed daily. If you live in the north, you will be chipping the ice off their water troughs too. It is a 7 days a week, 365 days a year existence. It can take some getting used to.

Also, keeping livestock can give you both incredible highs and desperate lows. Having lambs die in your arms is a heartbreaking experience. Even after being warned about this, we dismissed the warnings as negative thinking. But it happened. Burying a lamb is one of the hardest things I have ever had to do. At the high end, delivering a newborn lamb and watching it suckle for the first time is an experience that is hard to beat. Fortunately, there are quite a few amazing experiences you can have living this life.

The next few chapters in this book dive into more detail about the expenses and incomes of keeping animals. This will help you plan what you want to do and how you can make money from your smallholding.

23

SARKA

Sarka came to us in 2015 with her seven woolly friends. We had bought three lambs the year before, some of which were Sarka's offspring. One day, I spotted an advert in which the owner of Sarka was selling her flock of eight. In one of our trademark "act first, think later" decisions, we offered to buy them all. While it did give us problems with pasture, it was also instrumental in us moving to a larger smallholding. Looking back, it turned out to be serendipitous.

Of all our sheep, Sarka was the flightiest. A slight puff of wind or a passing dandelion seed would be enough to make her jump. Moving fields was always a bit fraught, and Sarka always ended up in a panic. It was not easy getting her into a pen when she needed treatment or vaccinating.

At the same time, Sarka was an incredibly beautiful sheep. While most Coloured Ryelands turn white over the years, bleached by the sun, Sarka retained her dark, sultry looks. Not only this, Sarka had long legs and a slim body giving her a gazelle-like grace and demeanour. Sarka's unusual looks set her head and shoulders apart from her fellow field companions. Further, she was always up on her hind legs nibbling on hawthorn branches along the hedgerow, something the other sheep didn't seem to do. We thought Sarka was just great and were most disappointed to be told one day that despite her striking looks and aptitude for hedgerow surfing she wouldn't win any prizes in a show. Ryeland sheep are naturally squat and barrel-shaped, so sadly there wouldn't even have been any point in Sarka entering a show.

Of course, she was the first of our sheep to get fly strike. At that time, we used a spray to protect them from the blowfly, but, as we found out, it was only partially effective. One day, Sarka emerged from a pig arc as though she'd just seen a ghost. She literally flew out covering a huge distance in a single leap while looking over her shoulder to try and identify what was attacking her.

We had never had a case of fly strike before, so at first we didn't know what was wrong, but we set off to try and catch her so we could examine her. For once, she didn't make it too hard. Perhaps she found us less of a worry than the "ghost". We found the source: some tiny maggots under her tail. We treated her, killing all the maggots, and in no time she was back to normal.

It was around this time that we switched from trying to round sheep up to training them to come to us. I say "we", but it was Nicole's mission really. I just benefited from the results. While most of the sheep responded well and became friendlier, Sarka remained suspicious.

Sarka was put to the tup with all our ewes in the autumn of 2015. The following spring, she delivered two lovely lambs, a boy and a girl. Sarka, like most of our sheep, needed a little help while giving birth. However, despite her skittish nature, she was happy to let us help. We saw this as a breakthrough.

In the summer of 2016 Sarka became unwell, so we had to separate her from the main flock. We brought her into the paddock by the house with another sheep, Sky, to keep her company. It seemed to be a bad case of post-pregnancy worms, but they were proving quite resistant to treatment. In the end we had to call the vet, who took a blood sample and advised a course of treatment which, thankfully, worked. The vet also mentioned that Sarka had a heart problem. During this time Sarka had a lot of handling, so, bit by bit, she became more trusting of us.

In the spring of 2017, we moved north to south-west Scotland. Sarka survived the journey and, remarkably, emerged from the trailer quite calm and contented. Perhaps it was the sight of all that fresh grass, or perhaps she had just enjoyed the journey, but there was no hint of her usual skittishness.

All our sheep settled in quickly; they lapped up the fresh

surroundings and increased acreage. Sarka seemed to have left her skittishness behind her. Despite there being no hedgerows to browse, she really blossomed.

During these first few weeks in Scotland, Nicole spent a lot of time with the flock making sure they were settled. During this time, Nicole and Sarka became very close. Sarka started leaning in for pats and head scratches and maybe, if she was lucky, a few sheep nuts. Sarka's transformation was a joy to behold.

Amazingly, Sarka's health seemed to improve too. Previously cursed with a slightly runny bottom (a magnet for blowflies), she dried up and had a really good summer. Her condition was the best it had ever been.

Shearing is always a little stressful for sheep – and for Sarka, in particular. That summer, after shearing was complete, we found Sarka panting and her heart racing. It felt different from her "normal" panic attacks, so we called the vet for advice. The vet came and said these problems were down to Sarka's heart condition. She treated Sarka with a number of injections but told us the prognosis was not good. We kind of felt like the vet was hinting that we would be best putting Sarka down. We didn't consider that to be an option and instead purchased some of the medicines so we could treat Sarka quickly should this happen again.

Despite her prognosis, Sarka continued to thrive. We felt that the more mountainous terrain and space suited her. Also, our fields are away from roads, so there was little disturbance from strangers. While she loved Nicole to bits, it still took a bit longer for her to accept me, but she did in the end. It was a lovely feeling having Sarka press into me as I scratched her back.

As we moved from autumn into winter, tupping time came around again and we had a tricky decision to make. We had already decided not to put any of our ewes who'd had problems lambing in the past to the tup again. At the same time, we wanted to give a better chance to all the ewes we felt we had let down in the past.

Sarka was in neither category and was as healthy as she had ever been. The decision we faced was that the vet had advised against putting Sarka to the tup due to her medical condition.

Well, we had overruled the vet before and, in our usual fashion, we did so again, so Sarka was put to the tup. It all went well and Sarka was doing nicely. But then, one day, she had another attack of palpitations. We had to administer her medicine including a diuretic. Sadly, this caused Sarka to abort spontaneously. In hindsight, we should have listened to the vet.

As lambing approached, we needed to split the ewes into separate flocks of pregnant and non-pregnant ewes. The pregnant ewes required a high-protein supplement, and it was far easier to provide this without the others muscling in. Also, we wanted to avoid any problems during lambing through overcrowding or jealousy.

The sheep seemed quite happy with this arrangement. Well, all except Sarka, who was entirely unimpressed when she was not included in the group of mums. She would stand on the hill near the lambing paddock and bleat for all she was worth. We suspect we had separated her from her best sheep friend. It goes to show, sheep do have relationships amongst the flock and they don't like to be separated from those they like best.

Luckily, this didn't affect Sarka's relationship with Nicole, which continued to prosper.

24

SHEEP

About Sheep

It is my impression that most people tend to regard sheep as a bit of a simple animal. That couldn't be further from the truth. Once you get to know them, sheep are actually remarkable animals. They can be affectionate, playful and even cheeky at times. They are also, in their own way, quite intelligent. For example, they are better at remembering where to find gates than our so-called intelligent breeds of dog.

When we first got sheep, experienced farmers and smallholders alike used to tell us that from the moment sheep are born, they are just looking for ways to die. We thought that a bit pessimistic and harsh, but they were right. You need to keep a close eye on your sheep.

If you are thinking about getting sheep, the first thing to think about is what breed of sheep will be best for you and your smallholding. There are many breeds which have all been bred for different purposes. Some are hardy and bred to survive on the hills of Scotland, Wales and Northern England. Some were bred for wool. Some were bred for meat. There are all sorts. The best approach is to start with why you want sheep in the first place and work from there. Another thing to find out, if you are planning to breed and sell sheep, is which breeds are popular in your area. As we have discovered, the market for Coloured Ryeland lambs in south-west Scotland is quite small.

If you are thinking of showing your sheep at country shows, it is a good idea to choose a breed which is shown at shows near you, otherwise you may be in for some long journeys.

A bit of research goes a long way.

When we first decided to get sheep, we saw little further than them being lawnmowers. We did a little research and chose a breed recommended for smallholdings and those new to sheep. As a result, we have Coloured Ryelands; and we love them. As described, they have turned out to be friendly, playful and gentle, and happen to look like teddy bears.

Looking after your sheep will keep you on your toes. Like most animals, sheep can get into all sorts of bother. We have, more than once, found one of ours that has rolled onto its back and found itself unable to back up again. This can be fatal for sheep if they are not found and righted within hours. They have also, more often when young, got their heads stuck in fences.

They can also get ill and keep it secret from you until it is almost too late. Experience has taught us that a minimum of two checks a day is essential.

Aside from the domestic dog, the biggest threat to sheep is the blowfly. Given the chance, it will lay its eggs in the wool. When the maggots hatch, they start to burrow into the sheep and, basically, eat it alive. This is called fly strike. You can spray sheep with anti-blowfly sprays, but these are not 100 per cent effective. The problem is worse in lowland pastures, especially those with hedges nearby. Being able to spot the symptoms of fly strike is a key skill you need to develop as fast as you can. The earlier you catch it, the less harm done and the faster the recovery. You can, if vigilant, spot it before the maggots start burrowing. Their wriggly presence is enough to change a sheep's behaviour from subtle to significant.

So, all in all, sheep are a great addition to the smallholding. Just bear in mind that they are high maintenance, so will take up a lot of your time.

Specific Needs of Sheep

At a fundamental level, sheep's needs are simple. They require space to roam, grass and water. In winter, they will also need hay. They can also be quite partial to haylage, but, once open, haylage needs to be eaten within a week. If you are not careful, a lot can be wasted.

Beyond that, we have found some interesting things. Firstly, sheep like field shelters. While more than capable of surviving outdoors all year round, they will seek shelter from prolonged heavy rain. They can get wet through, much like our waterproof clothing eventually lets the water in. Once they get this wet, they can soon become depressed, ill and possibly even die.

They also like to get away from direct sun; some trees or a shelter can provide this. The dappled shade from trees is excellent on sunny days.

Another thing sheep like is space to roam. There is all sorts of information about how many sheep you can have per acre and about moving them regularly from field to field. What we have found is that when we give them access to all our pastures, they seize the opportunity to roam where they like. In fact, they seem to have meetings before heading off. The ability to choose where they go seems to be quite liberating for them.

If this is not possible, they do need to be moved regularly from field to field. This helps to keep the parasitic worm count down amongst other things. It also stops them getting bored. A bored sheep will look for ways to find new pastures on its own. They can be excellent escape artists. While our sheep are happy with the amount of pasture they have, it is common for us to find sheep from neighbouring farms in with ours.

There are often stray sheep on the road into town as well. They can, when they feel like it, scale stone dykes, especially those that have crumbled in places.

Sheep also need regular "maintenance". Specifically, they need their teats, feet and bottoms checked.

The underlying factor here is that these animals are being kept in what, to them, is an unnatural environment. It is like having dogs living in houses; it is not their true, natural environment. A side effect of this is that modern dogs need their toenails clipped. Wolves, the ancestor of the pet dog, do not.

The hooves of sheep are much like the nails on our hands and feet, so, as such, they grow continuously. In their natural environment, sheep would roam on rocky outcrops as well as grassy pastures. Walking on rocks would naturally keep the hooves in trim.

In grassy fields, the hooves can grow and cause foot problems, so one of your jobs will be to check hooves on a regular basis.

You should also keep the area around their bottoms trimmed. This is called "dagging". Mucky bottoms attract flies, especially the blowfly. By keeping the wool in this area short and clean, you can reduce the incidence of fly strike.

Finally, you will need to vaccinate your sheep annually along with monitoring, and treating them, for liver fluke and worms.

Handling Sheep

If you take yourself on a smallholding and/or sheep-handling course, it is likely you will be shown a sheep race. This is a long, steel passage about the width of a sheep with gates at either end. The idea is that you round up the sheep into a pen attached to the race and then drive them through. While in the race, you can examine them or administer medication. There are couple of problems with this approach, however.

In our early sheep owning days, having done such a course, we purchased a sheep race. Our first problem was where to put it. Ideally, it needs to be in a shed, secured, and close to the main house. Rural crime is on the rise as I write this. As we had no such facility, we assembled it in the field. The sides are large and heavy, so we left it there. Before long, it was stolen.

The second problem is the concept of rounding up sheep. Large-scale sheep farmers have dogs. These are highly trained working dogs and need to be working pretty much daily, otherwise they can suffer psychological problems. As a smallholder, you probably won't need such a dog. This means you will have to round up the sheep yourselves. The moment you enter the field with this in mind, the sheep will read your intentions and decide that they will be going anywhere except that pen. You can try all the tactics you like – for example, driving them along a fence into a large opening – but they will be having none of it. It just ends up with all of you, sheep and humans, stressed out and scattered.

That said, having the race stolen turned out to be one of the best things that happened to us. We changed our whole methodology from rounding up sheep to teaching the sheep to come to us. This took time, as sheep are naturally suspicious of humans. They see us as predators, so their natural instinct is to keep their distance. If you move towards them with intent, they can feel as though they are being hunted.

Our approach was to spend time with them and to tempt them in with sheep nuts. Once a few of the braver ones had worked out that it was rewarding to approach the humans, the others began to learn. It took time and a lot of patience but was worth every minute.

Nicole has our sheep so well trained now that she can call some of them into a pen by name. She did this one day when the vet had been called. The vet was looking a little concerned when she arrived to find all the sheep out in the field but none penned up. She looked on in amazement as Nicole walked over to the pen, called Peaches and watched Peaches run down the hill and straight into the pen without any of the other sheep following. These moments are priceless.

This approach means that we can enjoy our sheep more. At sheep checks, many will come over as they have learnt there might be back and head scratches. You get to know them individually. This makes it much easier to spot when things are not right. Early diagnosis of problems makes them much easier to deal with. It can mean the difference between one injection and months of repeat treatment.

Shearing

Sheep need to be sheared annually. Depending on where you are in the country, it can be tricky to find shearers willing to shear small flocks as it is not really worth their while. Some will try to fit small flocks in between large jobs. But, for most, shearing small flocks is more hassle than it is worth.

You can shear them yourself – but be warned, sheep shearing is a difficult job. It is very physical and extremely tiring. Like many things, it needs training and a lot of practice to get good at it.

It can also be a bit risky for the sheep. Shearing handsets are complex and powerful pieces of equipment that can easily cut a sheep. Nicole and I did the training, and we found it intensely rewarding. Having three sheep, we spent a fortune on sheep shearing equipment and set about shearing them ourselves. It took us about half an hour to shear each sheep. By the end of it, we were all hot, bothered and the sheep looked like they'd walked through a hedge backwards.

We came to the conclusion that we would have

professionals shear our sheep in future. The best shearers can shear a sheep in around a minute. Sometimes the sheep enjoy the experience. With an experienced shearer, the sheep is constantly moving, so they have no time to become uncomfortable. But even with experienced shearers, it is worth checking for cuts and nicks afterwards. That's when having sheep that trust you can be a godsend. You can just walk up to them, check them and treat any nicks, should you find any.

Legalities

If you keep sheep, you need to register with the relevant government agency. These differ across the different countries of the British Isles. For example, in England it is DEFRA, but in Scotland it is ScotEID. Before moving any animals onto your holding, you must have a CPH number. You will also need a flock number. You can find out how to do all these things online.

You need to keep detailed records of sheep movements on and off your holding. Details of all sheep movements need to be sent to the relevant authorities. Once you are registered, they'll send you instructions.

You need to keep detailed records of all medicines used along with batch numbers and withdrawal periods.

All the above used to be on paper using books supplied by the relevant agencies, but now a spreadsheet is fine.

There is always the chance of a random audit, so this information needs to be kept up to date.

Equipment

The following are some of the essential pieces of equipment you will need to acquire if you are planning to keep sheep.

Item	Description
Hayrack on wheels	These come in lengths of 4ft, 8ft and 10ft, and can handle up to around 6, 12 and 14 sheep feeding together.
Sheep hurdles	These come in various lengths from 4ft to 8ft. These are needed to create temporary pens for examining sheep. They are also used to make lambing pens. Highly versatile – a must-have.
Hayracks for hurdles	These clip onto hurdles and are used to provide hay to sheep in pens. Particularly useful if you have to keep a sheep in. It is also a must-have if lambing.
Sheep nut trough	You may need, if lambing, to provide a protein supplement. A trough makes this easier. Galvanised steel troughs are the most hygienic.
Buckets	You can never have enough buckets. These can be used for providing water to sheep in pens and feeding nuts to an individual sheep.
Medicines and needles	You will need to keep a supply of antibiotics in your fridge along with syringes and needles. Livestock vets expect you to be able to administer injections yourself. There are other potions and related equipment to consider. Worming and fluke treatment are usually administered using an oral drench. If you need to spray to keep blowflies off, you will need the spray gun.

The following are some of the optional pieces of

equipment you might be interested in.

Item	Description
Sheep race	While often recommended by sheep owners, we have found that we don't need one.
Headstock or yoke	While not essential, we have had instances where we have had to remove an infected tag from a sheep's ear and also one where a sheep needed an injection in its eyelid. A headstock can make this much easier.
Feed store	A metal box to keep the feed (sheep nuts) in. It has to be metal; otherwise, the rats will soon chew a way in. If you can keep the feed indoors, you won't need this.
Trailer	These come in all sizes. The chances are you will be moving sheep at some point, so it is probably a good idea to have a trailer that is big enough to transport all the sheep you need to in a single journey.

Infrastructure

As well as the equipment listed, there are some major pieces of infrastructure to consider.

Item	Description
Water trough(s)	If you have a property with no water in the fields, you will need to install these along with the associated pipework. They come in plastic or galvanised steel.

Item	Description
Field shelter	Our sheep really appreciate their field shelters. The size you'll need will depend on the number of sheep you have, but a good rule of thumb would be to allow two square metres per sheep.
Lambing shed	If you are planning to breed sheep, a lambing shed is pretty essential. See the section on lambing (below) for more details.

Sheep Profit and Loss

One of the main purposes of this book is to explore the balancing act between time, effort and money in smallholding life.

The following sections provide some indicative finances for owning sheep. Note: these are based on prices in 2020. They are rough indicators only, so should be viewed as such. The main goal is to give you a framework or methodology with which you can use to work out the potential costs and incomes from sheep.

Setting-up Costs

The following costs assume you are starting your flock from scratch. The table gives indicative costs of each piece of equipment and a guide showing what you would need to buy for a starter flock. Ideally, you could set up such a spreadsheet yourself to calculate your own start-up costs.

The most variable cost is the price of lambs. Depending on the breed and market conditions, they could cost anything from £50 to £300.

Item	Price	Quantity	Flock of 5
Lambs (average purchase cost)	£100.00	5	£500.00
Sheep hurdles	£23.00	8	£184.00
Hay rack on wheels	£255.00	1	£255.00
Double-sided saddle-type sheep hayrack	£15.00	2	£30.00
buckets	£5.00	2	£10.00
Tagging equipment	£25.00		
Food trough	£42.00		
Water trough	£140.00		
	Total		**£979.00**

The above assumes you already have some of the infrastructure in place; if this is not the case, you can add in the relevant costs. It also does not include the cost of large-scale infrastructure such as fencing and field shelters. These are too variable in size and cost to portray accurately. As a rough idea, a small field shelter suitable for three or four sheep could cost around £400 to £500. The wood alone for our lambing shed, which doubles as a field shelter, was over £1,000. We saved a bit of money by building it ourselves.

Even without those, as you can see, setting up a small flock of five sheep is not going to give much change from £1,000.

It is worth looking for second-hand items. There are a few websites around that can help. We did manage to source some of the more common items such as hay feeders this way.

Annual Costs

Like all animals, sheep have a number of ongoing

expenses. Next is a sample spreadsheet showing most of the sheep-related expenses. These monetary values are based on our experience, but these will vary depending on your expertise and also your breed of sheep.

The table below shows just the cost of keeping sheep; it does not include breeding. Breeding (lambing) costs are provided later in the chapter.

	Per Sheep	Minimum
Vet	£38	
Medicines, syringes and needles	£11	
Drenches (excluding drench gun)		£119
Vaccines		£53
Anti-blowfly spray or pour on		£70
Shearing	£5	
Salt lick	£8	
Sheep nuts (for training)	£5	
Winter hay	£29	
Hidden costs	£14	
	£110.00	**£242.00**

So, each sheep will cost you around £110 a year, plus you will need to buy bottles of wormer, Heptavac (annual vaccination to protect against pasteurellosis pneumonia and clostridial diseases), anti-fly sprays and fluke treatment.

So, for example, a flock of 5 sheep would cost in the region, each year:

5 x £110 + £242 = £792

A flock of 10 sheep would cost in the region, each year:

10 x £110 + £242 = £1,342

So, as you can see, depending on the size of your flock, it can be quite an outlay.

An explanation of what each of the previous costs mean is provided here:

Vet One goal is to keep vet visits to a minimum. However, sheep can get into all sorts of problems. As you gain experience, you can diagnose and deal with many of these yourself. However, you might still find you are calling the vet more often than you had anticipated. We also ask the vet to analyse samples of sheep dung so we can administer targeted worming and fluke treatments. While overall more expensive than routine treatments (there is a fee to get faecal samples analysed), this approach helps avoid worm resistance and keeps the number of treatments down. It is also useful to know what else is living in your pasture.

Medicines, syringes and needles It is a good idea to keep a bottle of antibiotics in the fridge. This way, you can catch infections early and deal with them without recourse to the vet. You will also need an antibacterial foot spray, especially if your ground is predisposed to being wet, as sheep can be prone to foot infections.

Drenches	Depending on your approach to worming and fluke treatments, you may need to buy bottles of the relevant drench. Again, for small flocks, the vet may be able to make up individual treatments for you. For larger flocks, it is worth buying the bottle. You will need an appropriate drench gun. This can be kept and reused for years.
Vaccines	You will need to vaccinate your sheep with "Heptavac P" every spring. The smallest bottle contains 25 doses and, once opened, the unused vaccine must be thrown away.
Anti-blowfly spray or pour on	There are a variety of products available (at the time of writing, the two most popular are Clik and Crovect) that provide protection against blowfly strike on sheep and lambs. These products work in different ways, so it can be useful to ask for advice before buying them. It should be applied before an anticipated blowfly challenge.

Shearing	If you intend to shear your flock yourself, then you will have to factor in the cost of the shearing course and shearing equipment. Otherwise, you will need to find a shearer. For small flocks, they may also charge for the time getting to and from you. It is worth getting in touch with local sheep farmers to see if their shearers can fit you in while they are in the area.
Mineral supplements	You can provide minerals in various ways to your sheep. One is a salt lick: a solid block that they can lick as and when they need to. This needs to be set up under cover otherwise the rain can wash it away. You can also buy buckets containing mineral licks. These contain molasses, so can be very popular. They are meant to last for weeks, but, in reality, the sheep can lick them clean in only a few days. You can also get long-release boluses that you can administer using a specialised bolus applicator gun.
Sheep nuts	If your grass is good and you are not lambing, you should not need to give sheep nuts to your sheep. However, they are excellent training aids and also very useful for distracting sheep when administering injections. They can also be quite handy when you want to lead them from one field to another.

Winter hay	You will need to find a reliable source of winter hay. Hay bales come in various sizes, each with pros and cons. Small bale hay is the easiest to manage but is also quite rare, so it can be quite expensive. Large, round bales are more common now, but you will need machinery of some sort to move and transport them.
Hidden costs	There are a number of ways that sheep can cost you more money. We have a cupboard full of various potions such as organic anti-fly sprays and clay (kaolin) powder for helping unsettled stomachs. There are also items such as marker sprays, dagging tools, clippers, and so on. You may also need straw if you have field shelters, and don't forget the cost of fuel if you are using a quad bike to get around. These will be different for each holding, but it is worth thinking these through for your set-up. If you are planning to show your sheep, you may also want to get hold of the relevant grooming tools.

Income

Having looked at the costs of keeping and breeding sheep, the next question is, can you make sheep pay for themselves?

There are three basic routes to make money from sheep:

1. Breed and sell lambs
2. Breed and sell meat

3. Produce and sell wool or wool products

To follow are some rough guides to calculating the potential profits from these approaches.

Breed and sell lambs

Breeding sheep to produce lambs brings in a new set of costs.

Aside from the cost of a tup (ram), you will need extra feed (sheep nuts and mineral licks) as well as extra equipment. It is also likely that you will be calling out the vet. For some breeds of sheep, you may want to register the lambs with the breed society.

The costs of lambing will vary depending on both your level of skill at lambing and the ease with which your breed can give birth themselves. The more skilled you are and the easier the sheep give birth, the lower your vet costs will be. However, the ewes will need supplements, extra hay, high-protein ewe nuts and minerals.

The following tables are based on the following assumptions:

- A flock of 10 breeding ewes, annual cost £1,342 *(see earlier)*.
- A lambing rate of 18 lambs per 10 ewes and all survive.

Based on these assumptions, the rough cost per lamb is shown next:

	Best Case	Average Case
Vet		£36
Mineral licks	£24	£24
High-protein ewe nuts	£21	£21
Two vaccinations (Heptavac P)	£6	£6
Breed registration fees		£10
	£51	**£97**

Note: the above excludes specific lambing equipment of which there is a vast range, some of which is listed in the section on lambing to follow.

Also, I haven't included the cost of the tup; you might be able to borrow one.

Regarding income, if the lambs can sell for £100 each, the above figures show that your profit could be as little as £3. One dose of antibiotics and that profit is gone. You can increase this by keeping vet costs down and avoiding specialist breeds and related fees, but even then the profit per lamb would be, at most, £49. It is also worth bearing in mind that lamb prices are highly variable and breeds which are more commercial fetch lower prices.

The next thing to look at is that the above costs are purely the costs of the lambs themselves. In reality, you have to pay to keep your breeding flock. Once you factor in the cost of keeping the mothers (see earlier in this chapter in the Annual Costs section), the picture looks even more stark.

The cost of the ewes can be split over each lamb, so this would add an additional (rough) cost per lamb of £74 (£1,342 divided by 18) and our table now looks like:

	Best Case	Average Case
Vet		£36
Mineral licks	£24	£24
High-protein sheep nuts	£21	£21
Vaccination (Heptavac P)	£6	£6
Breed registration fees		£10
Annual cost of ewes	£74	£74
	£125	**£171**

So, the total cost of producing lambs could be somewhere between £125 and £171. You would need to sell them at these prices just to break even.

This does not take into account that you will have trouble selling the ram lambs, even if they are prizewinners at shows.

Taking all of these figures into account, you can see that it is hard to make any money from selling lambs at small scales. It is, in fact, all too easy to make a loss.

The bottom line is that at these scales it is almost impossible to make money, let alone a living, breeding and selling sheep. Even if you kept costs right down and could sell the lambs for £200 each, you would need to sell over 20 lambs just to cover the council tax.

This all may sound a bit pessimistic. It is not meant to put you off, far from it. What I hope to do is make you more aware of the potential costs of keeping sheep as a smallholder.

Breed sheep for meat to sell

There is a small but healthy market in the UK for high-quality, organic meat. Note: to get the "organic" label, you need to pay some serious money to the Soil Association or other organic body for certification. For most smallholders, the costs are prohibitive. What you can do is sell your meat with a description of its provenance.

You can sell the meat as lamb, hogget or mutton. Lamb is from sheep under a year old, hogget is from a hog (a sheep that is between one and two years old) and mutton comes from sheep over two years old.

The laws in the UK are very strict regarding the sale of meat. If you want to sell the meat, you must take it to a licensed abattoir and butcher. Please note: these laws are under constant review and change periodically, so it is always worth checking this.

Taking all the costs of breeding and keeping sheep and adding the costs of producing meat gives us a table along the following lines. Note: it shows the costs for both the average and best-case lambing costs from the previous table:

	Best-Case Lambing Costs		
	Lamb	Hogget	Mutton
Kg of meat per animal (approximately)	15	18	22
Cost of breeding your lamb (best case above)	£125	£125	£125
Winter feed		£29	£56
Slaughter	£40	£40	£40
Butcher	£22	£26	£32
Total Cost	**£187**	**£220**	**£253**
Price per kilo to break even	**£12.46**	**£12.22**	**£11.49**

	Average-Case Lambing Costs		
	Lamb	Hogget	Mutton
Kg of meat per animal (approximately)	15	18	22
Cost of breeding your lamb (best case above)	£161*	£161*	£161*
Winter feed		£29	£56
Slaughter	£40	£40	£40
Butcher	£22	£26	£32
Total Cost	**£223**	**£256**	**£289**
Price per kilo to break even	**£14.86**	**£14.22**	**£13.13**

– excludes breed registration costs

At today's prices, hogget and mutton of this quality is available at around £10 per kilo, so the above table shows just how difficult it is to make any profit from producing and selling meat. In fact, the above figures show it is more likely to be a loss.

Even if you can get your costs down, the meat can be hard to sell. The market for premium quality meat is small and there is plenty of competition.

Breed sheep for meat for your own consumption

Another option is to produce meat for your own consumption. While there are legalities to be checked, presently you are allowed to slaughter and butcher your own animals for your own consumption. In this case, you can offset the cost of producing your meat against what it would cost to buy. In other words, does it save you money?

In producing your own meat, the real advantage is that you know where the meat has come from. Most meat in shops comes from intensive or factory farms because that is the only way meat can be produced at supermarket prices. Producing your own is making a small but real statement against factory farming. It is a good feeling.

The disadvantage is that there's only so much meat you can eat, so there's only so much money you can save. Nevertheless, using the same assumptions as detailed earlier, the cost of producing your own meat is illustrated below:

	Best-Case Lambing Costs		
	Lamb	**Hogget**	**Mutton**
Kg of meat per animal (approximately)	15	18	22
Cost of breeding your lamb	£125	£125	£125
Sheep nuts (to bring on lambs)	£10		
Winter feed		£29	£56
Total Cost	**£135**	**£154**	**£181**
Cost per kilo of your meat	**£8.99**	**£8.55**	**£8.22**

	Average-Case Lambing Costs		
	Lamb	**Hogget**	**Mutton**
Kg of meat per animal (approximately)	15	18	22
Cost of breeding your lamb	£161	£161	£161
Sheep nuts (to bring on lambs)	£10		
Winter feed		£29	£56
Total Cost	**£171**	**£190**	**£217**

Cost per kilo of your meat   £11.39   £10.55   £9.86

This is a real indication of the delicate balance between time, effort and money. The previous table shows that for all the unpaid time and effort you have put in to raise these lambs, it might still cost you more to produce your own meat than placing an order with an ethical organic farm for the same produce.

Produce and Sell Wool or Wool Products

When sheared, the wool comes off in the form of a fleece. These days, you will be lucky to get £1 per fleece. For commercial farms, the money they get from the wool just about covers shearing expenses. For smallholdings, producing the wool actually costs you money.

The only way to make money from the fleece is to transform it into something.

Your first thoughts might be yarn. The number of balls of wool you can get per fleece will depend on the breed and quality of the wool. Let's assume you can get around 10 balls per fleece.

First thing to do is to spin the wool into yarn. You can either do this yourself or send it to a spinning mill. If you have it spun at a mill, it is going to cost you in the region of £5 to £6 per ball.

Given that commercial, mass-produced yarn is widely available at less than the cost of spinning yours at a mill, you can see the problem you face. While high-quality and natural, you will have to sell it at quite a high price.

If you can spin the yarn yourself, then you can keep your costs down. Just remember, you are not getting paid and you will also need to invest in some spinning equipment and maybe even some training. Nevertheless, this approach could net you a profit of £5 or so per ball of wool (depending on your price), so you could, theoretically, make in the region of £50 per sheep.

The other option is to turn the wool into high-value products. There are a multitude of things you can do, from knitting baby clothes to producing high-quality woollen rugs. Trying to describe the possibilities here would take an entire book in its own right. Suffice to say, if you can identify a niche product that is popular, there is the possibility of making an income.

Lambing

Springtime is welcome for so many reasons. One of these is the chance to see lambs frolicking in the fields. As they hop, skip, jump and play, they are a joy to behold. What is not seen is the monumental effort it has taken to get them there.

While the previous sections give some idea of the costs of breeding sheep, what they don't really go into is the time, effort and emotional cost of lambing.

Before I go any further, having lambed a number of times, I would say that lambing is one of the most amazing things I have ever done. But I walked into my first lambing with my eyes half shut. I thought it would be pretty easy. I was wrong.

Lambing takes a huge effort and a huge amount of time. It is weeks of interrupted sleep combined with sheer panic and euphoria in equal measures.

There are a number of things you can do to make it a bit easier. Top of the list is finding a mentor who can come and help at short notice. You can take all the lambing courses in the world, but in the middle of a crisis during the small hours of a freezing night in March, nothing beats having someone sitting next to you gently talking you through what you need to do.

For another, think about taking time off work. Aside from struggling to work in a sleep-deprived state, the more hours you are away, the more chance you will not catch a problem until it is too late. Lambs will die, and you will feel terrible. This again illustrates the balancing act we face in this smallholder life. By taking time off work, we stop earning money. By not taking time off work, we risk lambs dying. It is a hard choice.

Having good facilities really helps. Ideally, you need a shed with power, light and water. Watching a vet carrying out a caesarean at 4 a.m. by torchlight is not something you really want to go through. We have also found that having the lambing shed next to a small paddock is much better for the ewes. They seem to benefit from not being penned inside a shed, having access to grass as well as hay, and being able to mooch about a bit, sitting in the sun, etc.

Time spent preparing is time well spent. You will need all sorts of equipment and also antibiotics, syringes and needles close to hand. Equipment might include any or all of the following: a heat lamp, lamb warming box, lubrication, calcium deficiency medicine, twin lamb disease medicine, iodine, electrolyte, tags and tag applicator, tail/castration rings with applicator, thermometer, teats, bottles, tube feeder, prolapse harness, colostrum, milk replacement, steriliser, bottle rack or equivalent, the list goes on.

During lambing, work out a regular schedule of checks that works for you. The more often you check, the more likely you are to catch a problem early enough to be able to deal with it successfully. The "industry standard" is four-hourly checks. We prefer two-hourly checks.

Taking the time to make sure your lambs are suckling pays huge dividends. Lambs get everything they need from their mother's colostrum and are therefore less likely to fall ill later.

Finally, make time to enjoy your lambs. Go out and lie amongst them in the spring sunshine. Enjoy those moments when they clamber over you or leap onto your head like kittens. Watch them as they charge around the field bouncing off obstacles and pinging into the air as if on

springs.
 It is worth every second.

25

PIGS

About Pigs

When we first got pigs, we didn't really know what to expect. We'd heard all the usual things: they are very intelligent, can be very friendly, but will also eat you if you pass out in their run. We visited some pig keepers and talked about them, and it was all pretty positive.

Our first pigs came with bundles of personality and a right attitude. On the one hand they were playful and inquisitive, but they could scream for Britain if picked up. To be honest, at first, we were not sure what to make of them.

But, bit by bit they grew on us. In fact, it was not long before we realised that there is something special about pigs. You can't really pin it on anything in particular, but the combination of intelligence, curiosity and a natural desire to interact makes them an animal that you can grow incredibly fond of.

To have such a big, powerful animal rush over to see you can only bring a smile to your face. The magical feeling you get when you stroke a pig and watch it collapse in pleasure is second to none.

Of course, the problem is that you can grow very attached to pigs, so if you are thinking about rearing them for pork, it can be very hard to put them in the freezer.

Very hard indeed.

Specific Needs of Pigs

Pigs need space, a scratch post, food, water and a shelter with plenty of clean bedding. They also appreciate somewhere where they can make a wallow pool. As with most animals, the more space you can give them, the better. Based on pig runs we had visited, we thought we had built quite a large run for our pigs, but within three months we were having to extend it. We could see that they were getting bored. So, when setting aside space for pigs, set aside as much as you can.

Pigs are descended from wild boar, so they are essentially a free-roaming woodland animal. If you can offer them woodland, they will love it. It is not essential, but it does make for happy pigs. Also, they can be prone to getting sunburnt, so unless you are planning on getting out there with the sun cream, some shade would be a good idea.

Pigs also need a place to sleep. The most common of these is the pig arc. It is best to have a pig arc with a wooden floor so the pigs are not sleeping on the ground. The wooden floor will stay clean because pigs are clean animals. They set aside areas of their run as toilet areas and they don't foul their bed. In fact, they love fresh straw. It is a joy putting new straw into the arc and watching them rummage around in it making their beds. You will need plenty of straw.

Their food does need careful planning. If they have too much to eat, they will put on a lot of fat. This may not be a problem if they are being kept for breeding, but if they are kept for meat, then too much food will mean more fat than meat on your chops. The most available food is commercial pig feed which you can source from country stores. It does the job but, if you can source fruit and vegetables, so much the better. You could grow some yourself, although you will need to grow a lot, or you could try to get hold of unsold vegetables thrown away from shops.

There are some restrictions on feeding food from the kitchen to pigs. When I was at school, the leftovers from our school dinners went to pig farms. I am not sure that's allowed any more, but it is worth checking.

Apart from that, like all animals, pigs need access to a clean supply of drinking water.

Handling Pigs

Aside from giving them back rubs, you shouldn't really need to handle pigs except for tagging them and putting them into a trailer. That said, if they get sick, then you might have to give them injections.

As with all our animals, we try to establish a good relationship so the animals come to trust us. This is essential with pigs as, given their size and strength, you will struggle to get them to do anything they don't want to do.

When they first arrive, it can be a good idea to put them into the pig arc and lock them in overnight. This way, they will start to build a nest which will help them feel at home. Weaners are incredibly quick and nimble and will escape given half a chance. As we found out, once they have escaped, they are hard to catch. An electric fence will not keep them in (see the Infrastructure section later in this chapter) unless they have been trained to respect it.

If they don't come already tagged, the best advice is to tag them as soon as you can. Even piglets are quite strong and incredibly wriggly, so they don't make it easy. It is also worth tattooing them with your herd number. This is quite easy, you just buy the slapper with your herd number on it, dip it in some ink and a quick flick of the wrist does the rest. Slappers come with detailed instructions.

Giving pigs an injection can be something of a challenge. If they are under the weather, they'll probably just let you get on with it. But if, for example, they are put on a course of antibiotics and start to feel better, then you will be faced with injecting a pig that is not in the mood to comply.

There are no hurdles for pigs (they would just send them flying). Nor is there the equivalent of the cattle crush. In fact, the only thing on offer is a pig board: a flat board with which you are, allegedly, able to trap the pig in a corner. Good luck with that is all I can say.

If you are planning to put the pigs into a trailer, the best thing to do is train them first. Basically, bring the rear of the trailer up to the pigpen, open the door and leave it there. After a bit, throw in some tasty titbits like bananas or apples. They'll soon be running in and out. If you don't do this in advance, then all I can suggest is that you allow plenty of time to load them.

Legal Stuff

If you keep pigs, you need to register with the relevant government agency. These differ across the different countries of the British Isles. For example, in England it is DEFRA, but in Scotland it is ScotEID. Before moving any animals onto your holding, you must have a CPH number. You will also need a herd number. You can find out how to do all these things online.

You need to keep detailed records of pig movements on and off your holding. Details of all pig movements need to be sent to the relevant authorities. Once you are registered, they'll send you instructions.

You need to keep detailed records of all medicines used along with batch numbers and withdrawal periods. A spreadsheet is fine.

There is always the chance of a random audit, so this information needs to be kept up to date.

Equipment

The following are some of the essential pieces of equipment you will need to acquire if you are planning to keep pigs.

Item	Description
Pig arc	A pig arc, or similar, is essential for pigs. These come in various sizes, so you can choose one depending on the size of your pigs, whether you plan to breed, and so on.
Straw	Pigs need lots of straw for their bedding, so you'll need to give them fresh straw regularly.
Food trough	Galvanised steel troughs are the most hygienic.
Tags and tag applicator	Piglets can arrive with just an ink mark on them, so you will have to tag them.
Tattoo kit	You may, depending on your location and breed society rules (if applicable), need to tattoo your herd number onto your pigs. Kits are available for this.
Water trough	If you have a property with no water in the fields, you will need to install these along with the associated pipework. They come in plastic or galvanised steel.

The following are some of the optional pieces of equipment you might be interested in.

Item	Description
Pig boards	Robust boards to help you manage your pigs.
Feed store	A metal box to keep the feed in. It has to be metal, otherwise the rats will soon chew a way in. If you can keep the feed indoors, you won't need this.
Electric fence	This come in various shapes or sizes. It is best to get the ones specifically designed for pigs. You can get mains, battery or even solar-powered versions of these.

Infrastructure

As mentioned, you will need a pig run with a shelter, most likely a pig arc. The run can be fenced with a traditional stock fence or an electric fence.

Electric fences can work, but you first need to train the pigs to respect it. The natural reaction for a pig, on getting a shock, is to move forwards. So, the untrained pig will touch the electric fence with its nose, get a shock and then drive forwards to get away, and in the process demolish the electric fence. Training involves placing a barrier behind the electric fence blocking them from going forward so that they learn to go backwards.

A stock fence is also fine, but you will need to put barbed wire strands around the base. While animals like cows will try to get over a fence, pigs will look to get their noses under it and lift it.

Whatever you choose, it needs to be secure from the

moment the pigs arrive. Otherwise, they will zero in on any weakness and be off.

If you have a gate, it is also a good idea to have the top hinge facing down, otherwise the pigs can just lift the gate off its hinges.

Pigs Profit and Loss

One of the main purposes of this book is to explore the balancing act between time, effort and money in smallholding life.

The following sections provide some indicative finances for owning pigs. Note: these are based on prices in 2020. They are rough indicators only and should be viewed as such. The main goal is to give you a framework or methodology with which you can work out the potential costs and incomes from pigs.

Setting-up Costs

The following costs assume you are starting your litter from scratch. The table gives indicative costs of each piece of equipment and a guide to what you would need to buy for two pigs. Ideally, you could set up such a spreadsheet yourself to calculate your own start-up costs.

The most variable cost is the cost of weaners. Depending on the location, breed and market conditions, they could cost anything from £30 upwards.

Item	Price	Quantity	Two pigs
Piglets (weaners)	£40.00	2	£80.00
Tagging equipment	£25.00	1	£25.00
Tattoo Slapper and ink	£45.00	1	£45.00
Pig arc	£500.00	1	£500.00
Fencing	£300.00	1	£300.00
Food trough	£42.00	1	£42.00
Water trough	£140.00	1	£140.00
			£1,132.00

Note: fencing costs will depend on the size of the pig run and the type of fence used. The £300 above was the approximate cost of fencing off a 25m x 25m area with a 4-foot steel gate.

Annual Costs

Like all animals, pigs have a number of ongoing expenses. To follow is a sample spreadsheet with most of the pig-related expenses. These monetary values are based on our experience, but these can vary depending on your expertise and also your breed of pig. They will also depend on the type and quality of your pig run. The more space they have, the more natural foraging they can do.

	Per pig per month
Vet	£1.50
Food	£13.33
Straw	£8.00
Hidden costs	£1.00
	£23.83

This shows it costs around £24 per month to keep a pig. Note: the capital costs are not included in this table mainly for simplicity, but also because these will vary depending on your individual situation. If we were to include capital costs and assuming £1,052 was spent setting up the pig facility, then spread over 10 years the cost would be a further £8.77 per month.

An explanation of what each of the previous costs relates to is given below:

Vet	Our goal is to keep vet visits to a minimum. Pigs are pretty hardy and much lower maintenance than, for example, sheep, so it is probably OK to forecast minimal or even no vet costs. The previous table assumes one in five sows will get ill once per year with a vet visit costing around £90.
Food	Cost of buying pig food.
Straw	Assume two small bales per month per pig.
Hidden costs	There are ways that pigs can cost more money. But aside from paying for electricity to power electric fences, there are few other expenses. For us, it was mainly bananas – they just loved bananas, so we just kept buying them.

Income

The question is, can you make pigs pay for themselves? There are two basic routes to make money from pigs:

1. Breed and sell piglets
2. Rear pigs for meat

To follow are some rough guides to calculating the

potential profits from these approaches.

Breed and sell piglets

The first thing to mention is that if you plan to breed pigs, you will need access to a boar. You will also need to look into how best to look after the boar, as they will need to be kept separate from the sows for large parts of the year.

For the purposes of this section, we shall assume that the boar costs roughly the same as a sow and that the pregnant sows incur no additional costs.

We shall also assume an average litter of 10 piglets, all of which survive.

Number of breeding sows	2	5	10
Number of piglets	20	50	100
Cost of mothers per year	£572.00	£1,430.00	£2,860.00
Cost of boar	£286.00	£286.00	£286.00
Total cost	**£858.00**	**£1,716.00**	**£3,146.00**
Income from sale of piglets	£900.00	£2,250.00	£4,500.00
Profit from piglets	£42.00	£534.00	£1,354.00
Sale price needed to break even on piglets	£42.90	£34.32	£31.46
Profit per piglet (sale price £45)	**£2.10**	**£10.68**	**£13.54**

The above shows that the more pigs you breed, the higher the profit. This is down to the cost of the boar. The more sows, the more widely the costs of the boar can be spread. It also shows why farming is moving away from traditional small farms into large-scale operations.

It also shows it is quite hard to make much money

breeding pigs. Indeed, you would need to sell around 130 piglets just to pay the council tax. Selling that many piglets would be a tall order. You would be straying into commercial farming territory, and it is likely commercial buyers would pay a lot less per piglet.

Also, part of the reason for choosing the smallholding way of life is to escape the corporate lifestyle, but the mass breeding of animals is pretty much a corporate activity these days.

Raise pigs for meat to sell

There is a small but healthy market in the UK for high-quality, organic meat. Note: to get the "organic" label, you need to pay some serious money to the Soil Association or other organic body for certification. For most smallholders, the costs are prohibitive. What you can do is sell your meat with a description of its provenance.

The laws in the UK are very strict regarding the sale of meat. If you want to sell the meat, you must take it to a licensed abattoir and butcher. Please note: these laws are under constant review and change periodically, so it is always worth checking this.

Assuming the purchase of two weaners that are kept for six months, we get a table along the following lines:

Kg of meat per animal (approximately)	80
Cost of keeping pig for 6 months	£183.00
Slaughter	£40.00
Butcher	£96.00
Total Cost	**£319.00**

Price per kilo to break even	£3.99

At today's prices, we can assume pork of this quality is

available at £7.50 per kilo. You might be able to increase this by creating pork specialities such as ham, bacon and sausages.

Suffice to say, assuming some wastage (fat), the table below shows that you could make in the region of £225 per pig (excluding packaging and shipping costs). So, using our council tax example, you would need to sell the meat from six pigs just to pay the council tax. It might not sound much, but it is a lot of meat to sell.

Kilos to sell	64
Price per kg	£7.50
Cost per kg	£3.99
Profit per kg	£3.51
Total profit	£224.80

Rear pigs for meat for your own consumption

Another option is to produce pork for your own consumption. While there are legalities to be checked, presently you are allowed to slaughter and butcher your own animals for your own consumption. In this case, you can offset the cost of producing your meat against what it would cost to buy. In other words, does it save you money?

In producing your own pork, the real advantage is that you know where the meat has come from. Most meat in shops comes from intensive or factory farms because that is the only way meat can be produced at supermarket prices. Producing your own is making a small but real statement against factory farming. It is a good feeling.

Also, you can influence the taste by choosing what they eat. In some parts of Europe, pigs graze on acorns to get a particular flavour. Something to think about if you have a few oak trees handy.

The disadvantage is that there's only so much meat you

can eat, so there's only so much money you can save. Nevertheless, using the same assumptions as above and assuming you slaughter and butcher the pigs yourself, the cost of producing your own meat is illustrated below:

Kg of meat per animal (approximately)	80
Cost of keeping a pig for 6 months	£183.00
Cost of meat produced per kilo	£2.29

Of all the animals, this is probably the cheapest meat you can produce. Two pigs would provide a good portion of the meat for a family of four for around a year. That could be a decent enough saving on the shopping bill.

But it is not a living.

Breeding Pigs

The first thing to say is that breeding pigs can be a somewhat involved and risky undertaking. In commercial pig farms, the mortality rate can be as high as 40 per cent.

As a smallholder, you should be able to devote more time and energy to your sows, and then reduce this to the point where you can just about save all the wee piglets.

This reference to time and energy might sound familiar. In fact, once again you are faced with balancing your time, effort and money. Much as looking after your young piglets can be rewarding, it is basically unpaid.

The pig mating process is quite involved, so you'd do well to spend time researching and planning this.

Once pregnant, it is all about keeping your sows healthy. As with all breathing and living creatures, there is an army of pathogens just waiting to catch us out. The key thing is hygiene.

Little piglets need a warm environment. Taking time to plan and build a good farrowing pen will pay dividends.

This is a complex set-up, so it will be beneficial to carry out research and set it up properly.

Many books on breeding pigs talk about the risk of the sow crushing the piglets. The sow will need to get up and move about, so there is a risk if you have a lot of piglets scurrying around. The sow will be very careful – but remember, she's a big girl.

Raising your own pigs takes time, effort and commitment, as with raising any type of livestock. But, done well, breeding your own pigs can take you one step closer to self-sufficiency as well as giving you a lot of satisfaction.

26

CHICKENS

About Chickens

Chickens and a smallholding go together like bread and cheese. In return for food, board and lodging, chickens reward us with a steady supply of eggs. Such is their popularity that, in recent years, there has been an explosion of interest in keeping chickens in gardens in towns and cities. This has had the knock-on effect of making poultry supplies much more widespread; in other words, it is easy to find what you need in order to keep chickens.

There has also been an explosion of interest in what are called "rescue hens". Hens are most productive in terms of laying eggs in their first year. Commercial egg producers have historically sold off their laying hens after they reach one year old. These discarded hens would generally end up as pet food or something similar. But these days, through the charity British Hens Welfare Trust (BHWT), you can rescue these hens.

It is worth mentioning that chickens kept in a free-range environment do not lay every day. In fact, they are pretty seasonal when it comes to laying. After the autumn equinox, when the days start to shorten, it is common for laying to stop completely. It then restarts a few weeks after the winter solstice when the days are growing longer. This is why commercial egg producers prefer to keep the hens in sheds where they can control the lighting and "fool" the chickens into thinking it is always springtime.

Laying also stops at moulting – something chickens do from time to time. At moulting, their feathers start falling out and they can look a right state.

So, with egg production, it can be either feast or famine.

You might be wondering whether you need to get a cockerel. The short answer is that – unless you plan to breed hens – no, you don't need a cockerel. In fact, if you have a small number of hens, it is not that good an idea to get a cockerel. When cockerels mount the hens, which they do quite often, they leap on the hen's back and hold on with their large, sharp talons. If you only have a few hens, they will receive more attention from the cockerel than they can cope with and are likely to get wounded.

To hens, blood is like the proverbial red rag to a bull. If they see a hen bleeding, they can go into a frenzy and eat the poor hen alive. So, cockerels are best in larger flocks of, say, at least six hens. Also, cockerels are generally pretty big, so they need plenty of space. For a cockerel, you really need a free-range environment.

Having said all that, cockerels do have their uses. They look after the hens and call them over if they find food so they can eat first. They also keep an eye out for predators and warn the hens if they see something threatening nearby like a fox or bird of prey.

Specific Needs of Chickens

Like all animals, chickens need food, water, a house, a safe area and space to roam.

Having kept hens in both a town garden and in a rural setting, we have found that they really do best when given ample space to roam. If you do some research, you will probably come up with a number of around one chicken per one square metre in a chicken run. While chickens can live in such a set-up, what we have observed is that they really thrive when given much more space.

Ours have access to areas of grass, hedges and scrub. They use it all. Some days they spend mooching about under the hedge. On sunny days, they like to spread out sunbathing along the foot of the hedge or seek shade in the scrubby area. Ours even have access to the fields, but they don't really go there that often. They are happy with what they have, which is about 400 square metres for seven hens and a cockerel.

They also have their own field shelter as well as access to the lambing shed. Chickens don't have waterproof feathers, so they need to take shelter when it rains. It is worth noting that they don't take shelter where they go to roost, so they'll need somewhere else to hide from the rain.

Overall, what we have observed is that with ample room to roam, chickens are much healthier all round. For example, our chickens no longer have the sorts of problems with worms that they used to have when they were kept exclusively in a run. Whether this is down to the parasitic worm eggs in the ground being more spread out, the chickens being less stressed or Nicole's natural worming potions, it is hard to say. Probably all three. But we do think stress is an important factor. Although our chicken run in our previous house gave each chicken over two square metres of space, one chicken, Petal, was always hiding. When she came here, she blossomed and became a very confident hen.

Top of the must-have list for chickens is somewhere to have a dust bath. To be honest, it is a lovely experience to bring home rescue hens and see them finally realising that they can be hens. One of the first things they do is dig a dusty hole and then spend hours covering themselves in dust. It is like they are making up for lost time.

In the natural world, hens would roost in trees out of the reach of predators like foxes. Domesticated hens tend to roost in chicken coops. Most of these come with roosting bars that simulate tree branches. These coops are great because they keep the hens dry, provide shelter from winter elements and also provide areas in which the hens can lay eggs.

The most important thing with chicken coops is that they need to be open when the chickens wake up at sunrise. You can do this manually or you can invest in battery-powered automatic door opener and closers. If your hens are in a fox-proof run, then it is OK to leave the coop door open at nights. Otherwise, it needs to be closed.

I mention this because chickens do not like to be cooped up. If the door is not open at sunrise, they will become aggravated and can turn on each other. It only needs a hen to draw blood, and they get into a frenzy. So, you might find yourself opening the chicken coop to find carnage inside.

On the subject of predators, the number one predator is the fox. They are wily creatures who will find any weakness in your chicken set-up. Having lost two entire flocks to foxes, we are particularly careful about keeping our chickens safe. We have found the above-mentioned automatic door openers to be brilliant. We don't blame the foxes; it is what they do. We feel it is up to us to provide our hens with an environment safe from foxes.

Another thing we have discovered is that chicken coops, even when closed, are not necessarily badger proof. As a result, our chicken coops are sited within a fenced-off area with an anti-badger electric fence that comes on at night.

As with other animals, chickens can get ill or suffer from parasites such as worms, lice and mites. Red mites, arriving in the feathers of wild birds, can quickly infest a chicken coop. They come out at night to feast on the chickens' blood. If red mites are not dealt with quickly, the hens will soon lose condition and can abandon the chicken coop leaving them exposed to predators.

We have found that the best approach is to use plastic

chicken coops and dust them regularly with diatomaceous earth powder. Plastic chicken coops are much easier to disassemble and clean, so, if red mites do strike, you can simply take apart the coop and wash each section thoroughly. We have had wooden chicken coops, but both of these were burnt within a year as we found it impossible to control the red mites in them.

As ongoing prevention, we dust the chicken coop with Smite weekly and also mix it in with their bedding. This helps prevent the mites from getting to the chickens. This approach has worked well for us for a number of years.

Chickens can also be affected by parasitic worms, so regular worming is a good idea. You can buy chicken wormers, or, like us, you can use natural products such as crushed pumpkin seeds, garlic, and chilli powder.

Aside from that, chickens need minerals, especially calcium. Naturally, they would get calcium from eating invertebrates. If they are reliant on pellets for food, it is a good idea to supplement this with grit, which usually comes in the form of crushed shells.

We also give ours a mineral supplement based on some natural products. We have a routine of giving them corn in the morning and afternoon, so it is easy to add supplements to this routine.

One beneficial supplement for chickens is organic, raw cider vinegar. We add a few drops to their water. It helps their digestive systems to function well, aids against parasitic worms and is a good source of vitamins. It's also easy to make, assuming you have a source of apples.

Hens do occasionally become broody. This is more common in bantams, but any hen can suddenly decide it wants a family. If your hens are free-range, then the broody hen can simply vanish. They do like to find very secluded spots to nest.

If you have a cockerel, then you might even get chicks. I say "might" because in our experience hens sitting on eggs can be a bit hit and miss. We have had one hen produce one chick, another produce thirteen and at least four false alarms. As they say, don't count your chickens before they are hatched.

Having chicks born like this is great. For one, it is the most natural way and the chicks tend to grow up strong and healthy. It also avoids all the stress of having to introduce new laying hens as yours get older. However, the boys soon become a problem. Of the thirteen chicks hatched by one of our hens, nine were male. These were just as cute as the females until they reached what would probably be their equivalent of teenage years in humans. Overnight, they went from cute to rampaging thugs. As well as fighting each other, they ganged up on the females and basically terrorised them. You have to act. It is pretty much impossible to sell cockerels, so in the end ours went in the freezer. It is something you need to be prepared for.

Handling Chickens

Like most animals, with a bit of patience and some tasty treats, chickens can become quite tame. It is quite important to gain at least an element of trust as, occasionally, you might need to catch one to examine it or treat it. We have found that weeding draws chickens in like bees to honey, so this can be quite a good activity for bonding with your chickens.

If you do need to catch a hen, a good approach is to wait for dusk and then fetch the hen out of the chicken coop. Chickens become quite dopey in the dark. This is another time when plastic chicken coops come into their own as you can just take the roof off and reach in.

One of the most difficult challenges you can face when keeping chickens is introducing new hens. Hens really only lay regularly for the first two or three years. Commercial egg producers will sell off or cull the older hens every year. Ours enjoy a full life until they die naturally of old age.

From time to time, if they haven't produced chicks, you will need to buy in young hens. To say that hens do not take kindly to newcomers would be a huge understatement. In fact, it is war. If the chickens are kept in a run, you will need to divide it in two and let them live side by side but out of pecking range until they get to know each other better. If you have plenty of space, enough for bullied hens to escape, then it will go much more smoothly. In this case, you can just throw them in together and let them sort it out themselves.

Legalities

For small flocks, there is no requirement to register your flock with the authorities. If you are planning to keep a lot of chickens, it is worth checking the regulations in your area.

Due to the threat of avian flu, there are, from time to time, restrictions on moving chickens and also on allowing them to mix with wild birds. These are usually widely published.

Equipment

The following are some of the essential pieces of equipment you will need to acquire if you are planning to keep chickens.

Item	Description
Chicken coop	Chicken coops come in all shapes and sizes. We use plastic coops as they are easy to clean and excellent for keeping red mites in check.
Bedding	There are many types of bedding including shredded paper, but we have found that dust-free wood shavings work best for us.

Item	Description
Feeder	There are all sorts of feeders of all shapes and sizes. Despite what they say "on the box", most don't keep the food dry in the rain. They are also a magnet to the local bird and rodent population. It might be OK letting the local robin have a few pellets, but the food soon disappears when the crows find it. There are feeders which the hens open by stepping onto a plastic treadle. These work well, can be filled with a few days' worth of food, are vermin resistant and the food does not get wet and mouldy.
Drinker	There are many shapes and sizes of drinkers for you to pick and choose from.
Smite	Red mites are an endemic problem. Mixing Smite powder in their bedding is a good method of controlling red mites.

The following are some of the optional pieces of equipment you might be interested in.

Item	Description
Automatic door opener/closer	These are brilliant. You can set them to time or to detect dawn and sunset. They open and close the chicken coop door automatically.
Chicken run	These come in all shapes and sizes, or you could make your own. The size will depend on the number of chickens you want to keep. If you are out a lot, it might be worth looking at fully enclosed, fox-proof runs to keep your chickens safe.

Feed store	A metal box to keep the feed in. It needs to be metal, otherwise the rats will soon chew a way in. Alternatively, you can keep the feed indoors.
Poultry netting	Useful for fencing off areas and can be set up as an electric fence. Only keeps full-size chickens in.
Electric fence	If you have a problem with badgers, you can get purpose-built electric fences to deter them. These can be placed on a timer and set only to be on when the chickens are roosting.

There are other things you might need such as grit, wormers, mineral supplements and corn. It all depends on how you want to keep your chickens. Grit supplements are high in calcium, and hens need lots of calcium for their eggs. If they have access to plenty of land, they probably won't need a grit supplement. Corn is just one of a number of chicken treats you can buy to treat them and even teach them to become friendlier.

Infrastructure

A lot of infrastructure has already been discussed. In some respects, the problem with chickens is that there's so much choice. Chicken coops come in all shapes and sizes. The chicken run can be anything from a fully enclosed predator-proof chicken run to allowing your chickens to roam free range. How you choose to set up your chickens will very much depend on the size and shape of your holding.

Our run is simply bordered with a three-foot high stock fence. If they were so inclined, our chickens could easily fly over it, but they choose not to. It is probably not a solution for building a chicken run that you will find in any books about chickens.

All I would add here is that if you do allow your chickens to free range, you will need to fence off your vegetable patch. Chickens are very partial to broccoli.

Profit and Loss

The following are some indicative costs for owning chickens. These are based on prices in 2020. They are rough indicators only and should be viewed as such. The main goal is to give you a rough idea and methodology to work out the potential costs and incomes from chickens.

Setting-up Costs

The costs below assume you are starting from scratch. The table gives indicative costs of each piece of equipment and a guide to what you would need to buy in order to get started. Ideally, you could set up such a spreadsheet yourself to calculate your own start-up costs.

Item	Price	Quantity	Six hens
Chickens	£10.00	6	£60.00
Chicken coop	£440.00	1	£440.00
Chicken run	£385.00	1	£385.00
Door opener	£150.00	1	£150.00
Food trough	£42.00	1	£42.00
Drinker	£10.00	1	£10.00
			£1,087.00

You may not need all of the above, so you can amend the table to suit you. However, if you are starting from scratch, it can be expensive.

Annual Costs

Like all animals, chickens have a number of ongoing

expenses, but these are much lower than other farm animals. Below is a sample spreadsheet with most of the chicken-related expenses. These figures are based on our experience but can vary depending on your set-up.

	Per chicken per month
Vet	£0.67
Food	£1.00
Bedding	£1.00
Hidden costs	£0.50
	£3.17

This shows it costs around £3 per month to keep a chicken. Note: the capital costs are not included in this table mainly for simplicity, but also because these will vary depending on your individual situation. If we were to include capital costs and assuming £1,027 was spent setting up the chicken facilities, then over 10 years the cost would be a further £8.50 per month.

An explanation of what each of the above costs refers to is provided here:

Vet Chickens are relatively hardy, so it is probably OK to forecast minimal or even no vet costs. The table above assumes one in ten chickens will get ill once per year with a visit to the vet costing around £40. Note: if you are planning to get ex-battery hens, they can be much more fragile and illness prone.

Food This assumes each hen will consume 20kg of pellets each year. The more space they have, the less pellets they will eat, and vice versa.

| Bedding | Assumes a 100-litre bag of wood shavings that lasts three months and shared between six chickens. |
| Hidden costs | This includes treatments such as Smite and supplements such as grit and wormers. |

Income

The question is, can you make chickens pay for themselves? There are two basic routes to make money from chickens:

1. Breed and sell chickens
2. Sell eggs

To follow are some rough guides to calculating the potential profits from these approaches.

Breed and Sell Chickens

It is probably too hit and miss to rely on your hens getting broody and producing chicks on their own. If you are going to breed and sell, you will be better off investing in an incubator and chick-rearing facilities. Given that you have a cockerel, you can simply put fertilised eggs in the incubator and off you go.

It would be worth checking out your local area to make sure you have buyers for your chicks. You can't send chickens easily via courier, so sales will either be local people or by taking them to market. Also, it can be very difficult to sell cockerels, so you will need a plan for what you are going to do with the boys.

To give you some idea, the following table shows the possible prices you might be able to sell chicks for. It assumes that pullets for sale would be 16 weeks old and fully vaccinated. Note: for organic pullets, you will likely need certification from the relevant body, and that is not cheap.

Type of chicken / sale	Price	Number needed to pay council tax
Average price for organic pullet layer	£5.50	278
Average price for pullet layer	£4.20	364
Possible price to other smallholder for pullet layer	£10.00	153

So, using our council tax example, you would need to sell quite a few just to pay your council tax. For the cockerel chicks, you would most likely be selling those for meat. I am not sure of the legalities of this, especially the regulations regarding safe slaughter of chickens to sell as meat. If you are thinking about this, it's something you need to check carefully. The alternative is to try selling them (live) at a livestock market. Either way, given supermarkets currently sell chickens for under £3 each, you would be lucky to get much money, if anything, per chicken. Note: these estimates don't include the capital costs of the incubator and related equipment.

Sell Eggs

As with chicks, you need to have a ready market for your eggs. Around the countryside, you can find many honesty boxes and signs for free-range eggs. So, there is already plenty of competition.

If you are thinking about selling through shops, you will need to gain the relevant certification so you can mark your eggs accordingly. You also will need to ensure a constant and reliable supply. People eat eggs all year round. Hens do not lay eggs all year round. In the spring, we can't give our eggs away. In the autumn, we have to buy eggs ourselves.

This means you will need a large flock. Also, new egg-laying hens will need to be brought in annually to replace those hens whose best egg-laying days are past.

Free-range eggs sell these days for around £3 a dozen, so to pay the council tax you would be looking at selling 6,120 eggs each year. This is feasible, as you would actually need around 20 to 30 hens each laying 300 eggs a year. If you are selling through a shop, you will probably only get around £1 per dozen, so you would need to produce three times as many eggs and, most likely, keep up this level of production all year round.

27

COWS

About Cows

In some ways, cows are the "premier league" of farm animals. They are certainly one of the largest. Our experiences with cows have taught us that they are imposing, yet endearing animals with big personalities and are inclined to make mischief.

Like pigs and sheep, there are quite a number of different breeds, each bred for a specific purpose. Some, like Highland cows, are traditional breeds that have remained the same for centuries. Others such as Holstein and Jerseys have been bred as milk producers. There are also those, of course, like the Aberdeen Angus that have been bred for meat.

If you are thinking about cows, it is worth taking the time to get the right breed for you. Historically, small farms and crofts maybe kept one or two cows for milking. They were milked twice daily by hand just to provide milk for the family.

In this chapter I am assuming that, as a smallholder, you are looking to emulate those small farmers and crofters from years gone by and might be looking to keep one or two cows for milk, meat or even land management.

If you have decided to keep cows, a good first step is to survey your land. Cows are adept at escaping, so only the best stock fence or walls will suffice. Fences and walls will need barbed wire strands along the top to stop cows from leaning over and eating grass on the other side. If they do this, they will keep pushing forward and flatten the fence. Drystone dykes will also need barbed wire laid along the top for similar reasons. A cow is easily strong enough to knock a wall down.

Your pasture is equally important. Cows eat a lot, so you will need enough space. General thinking is to allow two acres per cow, but that will depend on the quality of the pasture. We have found that, as for the sheep, cows like to have access to a large area so they can roam freely. That said, they respond equally well to being moved regularly following a system known in farming as "rotational grazing". This is not dissimilar to rotating crops in your veggie patch and allowing areas to rest and recover. Whichever system you choose to follow, a lot will depend on the amount of pasture you have, other livestock you have on your patch, and the way you prefer to manage your pasture. Just be aware that if they are kept in a field for a period long enough to get bored, they will try to get out.

You also need to think about mud. It doesn't take much rain for the ground to become soft enough for cows to churn it into liquid mud. Areas surrounding feeders, water troughs and gates can soon become wellington-boot-sucking, impassable bogs of liquid mud. This can make life hard for you as well as for the cows and any other animals sharing the pasture.

Specific Needs

Cows need access to quality grass in the summer months. In winter they will require winter feed, which can be hay, haylage or silage. They also drink a lot of water, so you need a decent water system.

Some breeds need to be brought indoors to a cowshed or barn for the winter months. Historically, in Scottish crofts, cows were brought into the black houses that were the main residences at the time. These days, sheds or barns are more than adequate. Indoor quarters for cows need to be secure. You will also want to have relatively easy access so you can clean them out, preferably with access for a tractor if you want to make life easy for yourself.

Cows will need regular checking for and, if needed, treating for liver fluke and lice. Lice live in their coats and seem to be more active in the winter. For a healthy cow, they are more of a minor irritation. However, cows that are unwell or getting insufficient nutrition will develop more serious lice infestations. In some respects, the lice act as indicator that the cow is stressed.

Handling Cows

Our philosophy with cows is the same for all our animals: we like to work with them so they learn to trust us. The farming community is awash with stories of failed attempts to load cows into trailers, failed attempts to herd cows from one place to another, close shaves in barns, and the like.

Our Highland cows came with big, long horns and looked both majestic and imposing. Regardless, twice a day we'd be out there combing them; over time, they went from being suspicious of us to coming over for a scratch. We trained them to walk into the cattle crush willingly. Nicole became very adept at calling them. We used the winter supplement of cow nuts as an opportunity to scratch their heads so they learnt that it was pleasurable rather than threatening.

Although we used cow nuts to tempt them to come to us for a while, that path is fraught with difficulty. They soon come to expect nuts every time they see you and can get quite tetchy if you don't have any. The Highlands would use their horns to check out your pockets – by checking out, I mean ripping open. It can also be quite disconcerting to see a cow come hurtling toward you at full speed in their excitement. You just end up standing there hoping they will stop in time.

Having said all that, you do need a proper cow-handling area with cattle crush and pens. Cows, like any animal, may need treatment from time to time and they also require TB testing. They need to be restrained, and the only safe way to do this is using a cattle crush.

Legalities

If you keep cows, you need to register with the relevant government agency. These differ across the different countries of the British Isles. For example, in England it is DEFRA, but in Scotland it is ScotEID. Before moving any animals onto your holding, you must have a CPH number. You will also need a herd number. You can find out how to do all these things online.

You need to keep detailed records of cow movements on and off your holding. Details of all cow movements need to be sent to the relevant authorities. Once you are registered, they'll send you instructions.

You need to keep detailed records of all medicines used along with batch numbers and withdrawal periods. A spreadsheet is fine.

TB is prevalent in the UK, so cows are subject to routine TB tests. Your local animal health department will notify you about these.

There is always the chance of a random check, so this information needs to be kept up to date.

Equipment

The following are some of the essential pieces of equipment you will need to acquire if you are planning to keep cows.

Item	Description
Cattle hurdle	Ideal for making up temporary pens and also for loading cows into trailers.
Cattle crush	Absolutely essential if you need to examine or treat your cow. It keeps the cow immobilised so you or your vet can stay safe. These vary in cost depending on the features. Also, there are specialist crushes for horned cows.
Water trough	If you have a property with no water in the fields, you will need to install these along with the associated pipework. They come in plastic or galvanised steel, the latter being more robust.
Circular cattle feeder	Takes standard round bales of hay, haylage and silage.

The following are some of the optional pieces of equipment you might be interested in.

Item	Description
Tombstone circular cattle feeder	Variant of the circular cattle feeder made especially for cows with horns.
Straw	If your cows are overwintering inside, they'll need a lot of straw.
Food trough	For feeding them supplements. Galvanised steel troughs are the most hygienic.
Tags and tag applicator	If you are breeding, it is a legal requirement to tag every cow.

Item	Description
Feed store	A metal box to keep the feed in. It has to be metal, otherwise the rats will soon chew a way in. If you can keep the feed indoors, you won't need this.

Note: if you are going to be manoeuvring large bales of hay, straw and especially silage or haylage which are very heavy, you will need machinery. We use a compact tractor with spikes on the back.

Infrastructure

Aside from properly constructed fences and walls, the two most important pieces of infrastructure are the winter quarters and cattle-handling area.

Whether you need winter quarters will depend on the breed of cow you have chosen. That said, it is our feeling that all cows need access to shelter of some form during the winter months no matter how hardy they are. So, even if they are hardy creatures such as Highlands, you would be doing them a favour by providing a field shelter.

The handling area is essential. This is where you will bring the cows for examination, treatment and TB tests. The best solution is a permanent pen that funnels the cows into the crush. The crush itself needs to be secure as cows are large and strong. A concrete floor is ideal because it is something you can clean and disinfect – plus, it doesn't get churned up and muddy.

You can, if you need to keep the costs down, just buy the yoke. That's the front part of the crush, the bit of machinery that traps their necks. You can then use cattle hurdles as the box to the rear to keep the body from moving. It is not unknown for a cow to walk off with the yoke attached, so the yoke will need to be firmly attached to the ground.

Profit and Loss

The following are some indicative figures for owning cows. These are based on prices in 2020. They are rough indicators only and should be viewed as such. The main goal is to give you a rough idea and methodology to work out the potential costs and incomes from cows. The following figures are based on our experience with Highland cows. The cost of buying cows can vary from £100 to over £1,000 depending on age, breed and pedigree.

Setting-up Costs

The costs below assume you are starting from scratch. The table gives indicative costs of each piece of equipment and a guide to what you would need to buy to get started assuming you have nothing in place. Ideally, you would set up such a spreadsheet yourself to calculate your own start-up costs.

Item	Price	Quantity	Two Cows
Calves	£500.00	2	£1,000.00
Cattle hurdles	£110.00	8	£880.00
Cattle crush	£2,500.00	1	£2,500.00
Circular feeder	£450.00	1	£450.00
Tagging equipment	£25.00	1	£25.00
Food trough (for supplements such as cow nuts)	£42.00	1	£42.00
Water trough	£140.00	1	£140.00
			£5,037.00

So, cows are not a cheap option. You should also factor in all the labour needed to build their facilities. These costs assume you would be doing that yourself (for free).

In terms of the cost of setting up a cow overwintering shed, the assumption is that you will already have a shed of some form on your holding. If you are looking to build a new shed, you won't see much change from £10,000.

Annual Costs

Like all animals, cows have a few ongoing expenses. Below is a sample spreadsheet with most of the cow-related expenses. These monetary values are based on our experience, but these can vary depending on your expertise and your breed of cow. It assumes one vet visit per cow per year and that you will be overwintering them indoors.

	Per cow per month	Per cow per year
Vet	£7.50	£90.00
Winter food	£43.33	£520.00
Straw	£21.67	£260.00
Other costs (medication, etc.)	£8.33	£100.00
	£80.83	**£970.00**

This shows it costs around £80 per month to keep a cow. Note: the capital costs are not included in this table both for simplicity and because these will vary depending on your individual situation. If we were to include capital costs and assuming £4,037 was spent setting up the cow facilities, then over 10 years the cost would be a further £34 per month.

One way to get reduce these costs is to produce your own winter feed. That means cutting and baling your own grass. You could purchase or hire equipment to do this or engage another farmer to do it for you. The latter approach seems quite popular around here, even for larger farms. If you managed to secure a contract to produce your own winter feed for, say £10 per bale and had a hardy breed that could overwinter outside, the picture looks a lot better.

Taking out the cost of winter feed changes the table as follows:

	Per cow per month	Per cow per year
Vet	£7.50	£90.00
Winter food	£10.83	£130.00
Other costs (medication, etc.)	£8.33	£100.00
	£26.66	**£320.00**

An explanation of what each of the above costs means is given below:

Vet	TB testing is carried out by vets. You should also factor in the occasional visit for health problems. The table above assumes one vet visit per cow per year, with a vet visit costing around £90.
Winter food	This is based on current prices for round bale haylage.
Straw	If you will be keeping your cows indoors, you will need straw; if not, you can discount this expense.
Other costs	This includes medications such as anti-fluke and anti-mite pour-on solutions, plus some cow nuts for training purposes.

Income

The question is, can you make cows pay for themselves? There are two basic routes to make money from cows:

1. Breed and sell cows
2. Rear cows for meat

Note: I am staying clear of dairy farming as that is a scale of operations beyond smallholdings. If you are thinking about keeping some cows so you can produce your own milk, then you will have to use one of the following models as the milk comes from cows with calves, so you will have to breed.

To follow are some rough guides to calculating the potential costs and incomes from these approaches.

Breed and sell cows

To be clear, this book is not an instruction manual on how to breed and sell cows, as it is a specialist activity. The focus is more on how challenging it is to make money from breeding cows.

Having chosen a breed, you would need to check up on the specifics of the breed in terms of facilities needed for calving. For example, Highlands give birth outdoors and then hide their calves in long grass, whereas some other breeds are best brought indoors to calve. It is a good idea to have the cattle crush nearby should a caesarean be needed.

You will need tags, a tag applicator and perhaps some additional specialist equipment such as a calf puller, which is a sort of winch used when the calf is large and struggling to get out.

The following offers a rough illustration of costs. It assumes you will have a vet perform artificial insemination and that you will provide a high-protein supplement for around 10 weeks prior to birth.

High-protein supplement	£130.00
Artificial insemination	£50.00
Total cost	**£180.00**
Income from sale of calves	£500.00
Profit from calf	**£320.00**

This assumes a private sale. If you go to market, you will need to factor in market commission and transport costs.

If you factor in the annual cost of the mother cow, it wipes out all of the profit. Even assuming you have a hardy outdoor breed and make your own winter feed, the cost of keeping the mother cow will be £320 a year. So, you would need to fetch well over £500 per calf to make any profit.

Rear cows for meat

The following is a rough guide to the cost of producing your own beef. As previously mentioned, if you plan to sell meat, the regulations on slaughter and butchering are very strict. There is also some debate about the age you should put a cow to slaughter. The figures below assume the cow will be at least two years old. Note: this will mean you'll need to have enough summer pasture and winter feed for the calves as well as the mothers.

	grow own haylage	buy in haylage
Kg of meat per animal (approximately)	200	200
Cost of keeping mother	£320.00	£710.00
Cost of calf (vaccinations, etc.)	£100.00	£100.00
Cost of keeping calf for two years	£480.00	£1,065.00
Slaughter	£100.00	£100.00
Butcher	£300.00	£0.00

| Total Cost | £1,300.00 | £1,975.00 |

Price per kilo to break even   £6.50   £9.88

The cuts of beef vary a lot in price from shanks (cheap) to fillet steak (expensive). But it is a premium product, and you can find organic beef for sale for prices of around £15 per kilogram.

Being practical, if you managed to sell the beef for an average price of £12.50 per kilogram and sell all of it, that would generate a profit of £1,200. Using that as a guide, you would need to produce the meat from one and a quarter calves in order to pay your council tax.

If you have to buy in winter feed, the cost rises to £9.88 per kilo, so at a sale price of £12.50 per kilo you would make £525 profit. This means that the meat from three calves would pay your council tax.

As with all other animals, the key is finding a market for your produce. The moment you start looking at selling through shops, you will find your income slashed.

Breeding Cows

The basics of breeding cows are much the same as for any other animal. The keys are having the right knowledge, carrying out the preparation and having good facilities.

The biggest difference is cost. Cows, being larger animals, mean larger costs. So, it can be difficult to justify breeding cows when you can probably buy calves for much less than the cost of breeding them yourselves.

That said, if you are thinking about breeding your own cows, then, as with sheep, a mentor will be the best thing you can find. The mentor's knowledge will be invaluable but also, sometimes, it will take more than one of you to help a large calf into this world.

As is the case when breeding animals, if all goes well

there is not a lot for you to do. If the birth goes smoothly and the mother is a good mother, then all you will be called to do is keep an eye out for mother and calf.

If things go wrong, however, you will need to intervene.

With smaller animals, you can sometimes help them where they are or carry them back to a shed if you need to get them inside. For example, vets can carry out a caesarean operation out in the open where the cow stands.

With cows, it is not that simple. They are simply too big, which can make them potentially dangerous and impossible to move without some serious equipment like a large tractor.

In terms of birthing, the equipment you need will include a properly installed cattle crush and birthing aids such as a calving jack. You will need plenty of indoor space and properly constructed pens. This is all quite expensive.

As a smallholder, these costs can be prohibitive.

28

BEES

About Bees

Bees are intensely fascinating creatures. One of the joys of keeping bees is that as well as watching them forage amongst your flowers, you are able to watch them in the hive itself. Although they can be perturbed by the disturbance, you can still see them at work on the frames including those you have pulled out to inspect. It can be quite hypnotic watching them at work.

The main benefit of bees is that they fly around pollinating a wide range of plants. They are essential to us humans as, without them and their pollination, many of our crops would fail and our food production would be in peril. On top of that, they produce honey, a food coveted by humans and animals alike.

We have all heard that they can sting, and they certainly can. But honeybees sting as a last resort because once they have stung you, they die. Opening their hive is perceived as a threat, so they will try to sting you when you do this. Hence, all the protective clothing.

That said, there are bees…and there are bees. Our first bees were Buckfast bees, and they were incredibly tolerant of humans. I was able to mow around the hive with a power mower (bees hate vibration) and they just ignored me. One cold evening, we found some stragglers struggling to get back into the hive, but we were able to pick them up, warm them in our hands and then guide them into the hive.

Our second set of bees were the complete opposite. Opening the hive led to a swarm of angry bees buzzing around furiously.

It is all down to the queen and whatever genes she has inherited from her mating with the drones. This is why, if you are new to bees, it is a really good idea to source your bees from an excellent apiary. Such apiaries will be careful about their breeding methods and should be able to offer more peaceful breeds of bees.

You might want to find out if you are allergic to their stings before getting bees. I had been stung as a child with no reaction, but as an adult I found I had become allergic. Being allergic can make you a little nervous around bees, and that is not a good thing. Bees, like most creatures, can detect nervousness and it can make them feel threatened.

But, if bees are something you are interested in, it is worth investigating and seeing whether you can keep them. On a smallholding, there's bound to be somewhere they can go.

Specific Needs of Bees

What your bees really need is for you to know what you are doing. For this reason, if you are planning to keep bees, the best thing you can do is contact your local beekeeping association and get yourself on a training course. It is also an excellent idea to find yourself a mentor, someone who can come and visit your colonies and offer sound advice and help as needed. Between them, the association and mentor can help you with choosing the right hive and bees for your area.

In practice, bees need to have a decent hive and access to nearby water and pollen-laden plants. The siting of the hive is important, as a poorly selected site for your beehives can cause problems for your colonies over the winter. For example, it is not a good idea to put your hives in a frost pocket, an area prone to flooding, under trees or on the edge of a wood. Hives should also be protected from the prevailing winds and have the hive entrances facing south or south-westerly if possible.

On top of that, the entrances need to be sited so that the bees coming and going are not flying across pathways or other people's gardens. While the bees don't want to sting you, they can accidentally bump into you and sting you in their surprise. This happened to me the day before our wedding, leaving me with a nice, swollen face on the day Nicole and I were married.

After you have your installed bees, you do need to look after them. This revolves around weekly inspections during the summer months and treatment for the varroa mite. You also need, in the spring, to manage their propensity to swarm. Swarming is the bees' way of procreation, so the urge to swarm is strong. If they do swarm, the queen you spent all that money on will be off with the swarm leaving a new, unpredictable young queen behind.

You also need to be aware of what's going on around the hive. It is not uncommon for hives to come under attack from wasps, hornets and even other bees. An attack from any of these can wipe out your colony, so you need to be able to spot the signs and take the appropriate action.

Further afield, it is worth taking note of what sort of flora abounds. Fields of rape are hugely popular with bees as there are masses of small flowers close to each other. However, a rape crop nearby can cause a beekeeper's heart to sink. Bees feeding on rape can become tetchy and unpredictable. Rape seeds have hitherto been treated with neonicotinoids, an insecticide blamed in many quarters for causing damage to bees' nervous systems. With the EU imposing restrictions on neonicotinoids, if they are indeed the cause, then perhaps rape crop bee tetchiness will be a thing of the past.

On a more positive note, if you can see what plants your bees are foraging on, then you can relate it to the honey they produce. For example, apple blossom honey is quite different in taste and texture from ivy-based honey. Knowing where your honey comes from means you might be able to produce a premium version.

As with all creatures, bees are at risk from parasites and pathogens. The most common is the varroa mite, and your training will show you what to do about this. Otherwise, it is all about learning to read your bees and interpret what is going on from your observations of their behaviour.

From late autumn and during winter, you need to make sure they have enough food, especially if you have been taking honey from them. You can do this by leaving enough honey in the hive or buying in special winter bee food such as bee candy or fondant. Winter inspections are less frequent, but you still need to check, from time to time, to make sure that they are alive and they have enough to eat.

Handling Bees

Unlike the other animals in this book, bees are not receptive to training, so you can't really teach them to trust you or come over for back scratches. In fact, most of your interaction with bees will be disassembling their home to check on them or to steal their honey. If they do remember you, it is not likely to be fondly.

However, like all animals they respond to the energy you give off. The more calm, unflappable and almost meditative that you are, the less threatening you will come across to the bees. If you can keep yourself totally calm as they swarm angrily around you, then the chances are they will calm down more quickly.

The standard approach when opening the hive is to use a smoker. The idea is that the smoke calms the bees. It sounds simple, but there's a bit more to it than that. You have to get the smoke levels right, otherwise it can aggravate rather than calm the bees. This is why training and mentoring is so important. Having someone guide you on what to do helps you to learn quickly and this is better for everyone, both humans and bees.

Equipment

The following are some of the essential pieces of equipment you will need to acquire if you are planning to keep bees. It is far from an exhaustive list, but it is a good starting point.

Item	Description
Beehive	A home for the bees, these come in various sizes and shapes. In the UK, the most common are the National and WBC hives. It is a good idea to consult your local association and fit in with what local beekeepers are using. Based on our experience, we would recommend wooden hives.
Supers	Smaller frames added to the hive in which you encourage the bees to make honey.
Frames (also called foundation)	These are sheets of wax in a wooden frame upon which the bees build their hive. Bees create cells on these both to rear bees and store honey.
Protective clothing	A set of clothing, including gloves, that covers you from head to foot so you can avoid getting stung.
Queen excluder	A metal sheet with holes that allows worker bees, but not queen bees, to access the supers. This ensures that there are no brood cells amongst the honey.

Item	Description
Bee excluder	A wooden sheet with a one-way valve that allows bees out of the supers but not back in (so you can take the supers off with no bees in them to collect the honey).
Hive tools	A range of specialist tools for working with beehives.
Smoker and fuel	A device used to help distract and calm the bees during inspection.
Contact feeder	A plastic bucket with a gauze lid that enables you to feed syrup to your bees.
Wasp excluder	A piece of wood that you can use to reduce the size of the entrance in order to help the bees keeps wasps out, should wasps attack the hive.
Cover boards	A wooden sheet that sits between the top of the hive and the roof.

The following are some of the optional pieces of equipment you might be interested in.

Item	Description
Dummy board	This is a frame but with a wooden rather than wax insert. It is placed at the end of the brood chamber instead of a wax frame. It makes it easier to get frames in and out.
Entrance block and wire mesh lid	A block of wood to put in the entrance and a wire mesh sheet that replaces the roof. These are used when moving hives. You need the wire mesh so the hive can get air (having blocked the door).

Item	Description
Ratchet strap	For keeping the hive together when moving it.
Honey jars, lids and labels	For putting honey in.
Honey extractor	A device for extracting honey.
Honey bucket	Something to extract the honey into.
Honey strainer	A sieve for filtering bits and pieces out of the honey.
Honey extraction tools	The super frames need to be uncapped in order for the honey to flow, so you need tools to do this.

Profit and Loss

The following are some indicative figures for owning bees. These are based on prices in 2020. They are rough indicators only and should be viewed as such. The main goal is to give you a rough idea and methodology to work out the potential costs and incomes from bees.

Setting-up Costs

To follow is a table showing the indicative costs of setting up a beehive from scratch. It assumes you will have to buy everything. There is a chance you may be able to source some equipment or bees for free via your local beekeeping association, and that would certainly help. The following is intended as a framework that you can use to calculate your own costs.

Item	Price	Quantity	Total
Starter Kit (contains all you need except bees)	£600.00	1	£600.00
WHBC Hive	£366.00		
Nucleus	£250.00	1	£250.00
Frames	£20.00		
Foundation (pack of 10)	£5.00		
Smoker	£35.00		
Hive tools	£15.00		
Honey extractor	£400.00		
Protective suit	£90.00		
Protective gloves	£15.00		
			£850.00

The above is just a single hive plus bees with equipment for one person. The starter kit will contain most of what you need for the bees, but none of the honey extraction equipment. You may need to purchase additional suits for other members of the family. Also, you will most likely find that you'll need additional hives sooner than you think. Ideally, you will need at least one extra in case you find yourself having to house a swarm. Once you are known as a beekeeper, you will soon be getting phone calls from nearby homes with swarms in their gardens.

Set-up costs for bees are not cheap. As mentioned, this can be mitigated if you get to know local beekeepers, as some may be able to lend or even give you some equipment. You may also, if you are lucky, be able to source a swarm from a local beekeeper. We, on occasion, were able to supply the odd swarm to local beekeepers to help them out.

Annual Costs

There are few costs associated with keeping bees once they are up and running. You will need to purchase some winter feed, especially if you are extracting honey given that honey is the bees' winter store. Aside from that, you may need to replace bits and pieces here and there, especially frames, foundation and smoker fuel.

Other likely costs you will face include buying hives or nucleus boxes to house random swarms you will find coming your way.

Income

There are three routes to make money from bees:

1. Sell honey
2. Sell bee-related products
3. Sell bees

To follow are some rough guides to calculating the potential costs and incomes from these approaches.

It is worth noting that all the costs assume that your hives remain healthy and strong. There are a lot of problems with beehives worldwide at the moment. There are numerous stories of hives collapsing, bees disappearing and the theft of hives. European bees are still struggling to come to terms with the varroa mite. It is possible that you will lose some of your bees and, therefore, face replacement costs. These could be, for example, the cost of buying new bees or the operational costs of taking no honey while you let a new colony get up to strength.

It is hard to estimate this sort of cost as it is, by nature, random, so it is not included in the following tables. If you are contemplating making money from bees, it would be a good idea to factor some contingency into your forecasts.

Sell Honey

This is the most popular route to make money from bees. There can be a strong demand for locally produced honey. One of the problems you will face is that the amount of honey produced each year will vary. Bear in mind that it is not a good idea to take honey in the first year of a hive; you need to give the nucleus time to develop and grow strong.

For a fully functioning hive, the general amount of honey produced in a single year would be between 8kg and 27kg (20lb and 60lb). The costs and income for this range is illustrated in the table below. These are general estimates.

Amount of honey produced	8kg	18kg	27kg
Number of jars produced per hive	40	80	120
Cost of replacing super frames, winter feed, etc.	£25	£25	£25
Cost of jars and labels	£24	£40	£64
Replace equipment every 15 years	£56	£56	£56
Total Cost	£105	£121	£145
Income (72 jars per hive at £4.50 each)	£180	£360	£540
Profit	£75	£239	£395
Profit per jar	£1.88	£2.98	£3.29

So, in a good year a hive could generate an income of £395. Using this figure, you would need to sell 460 jars of honey produced by 4 hives to pay your annual council tax. In an average year, this increases to selling 500 jars produced by 7 hives. In a bad year, it further increases to selling 800 jars of honey from 20 hives.

A key question is, how many hives do you want to keep? As 20 is a lot, once again you face a balancing act of the amount of work needed against how much money you hope to make. If you are thinking of making a living from honey, you would be looking at scaling up to managing a lot of hives.

As with all the other produce, you need to find a market for your honey. Depending on the year, 460 to 800 jars of honey is quite a lot to sell. Don't forget, you will have competition. Go along to any local market and you will always find people selling honey. Also, there are many signs on roads advertising honey for sale. It is possible to succeed with honey, but it will require time and effort.

Sell bee-related products

Some of the by-products of keeping bees can be turned into saleable items. For example, beeswax can be made into a number of items such as candles, polish and cosmetics. There is also propolis and royal jelly, both of which have medicinal qualities.

If you are interested in producing such items, then, as with everything, it would be a good idea to work out what you can sell them for and where your market is. As these are by-products of something you are already doing, the only real cost is your time. The amount of wax, propolis and royal jelly you can produce is almost impossible to predict, so financial tables are not offered here. If you choose to go down this route, you will probably need to proceed on a trial and error basis.

Sell Bees

There is a steady market for bees, both nuclei and individual queens. This is altogether more complicated, as any bees you sell need to be inspected and disease free. Also, you really need to know what you are doing when breeding queens.

Queens take on the characteristics of their own personalities plus those of the drones they mate with. Left to herself, a new queen will fly out one day into a swarm of drones which may have come from miles around. You cannot predict their temperament, so you can't just let your bees swarm and hope for the best.

Breeding and selling bees is an area of niche expertise, therefore it is not covered in this book.

29

SELENE

I have never forgotten the spring of 2017. The weather was foul; it just never seemed to stop raining. The 11 of us were in a fairly small field with our feet permanently under water. It was all too much for Peaches – sadly, she lost her lambs. It was tragic, but out of every tragedy some good can emerge.

In this case, all the ewes were brought into a shed. At first it seemed quite small. After all, there were 11 of us, but the prospect of clean and dry conditions overcame any reservations we might have had. Clean straw, plenty of hay, fresh water and, best of all, we were dry. We would lie there chewing cud and watching the rain falling over there.

Best of all, every day we had a treat – sheep nuts. We love sheep nuts. If you are a human reading this, think chocolate. In fact, when one of the humans brought them, we were so excited that we swarmed around them en masse and sometimes even trapped them up against the wall. Fortunately, they were able to extricate themselves and pour the nuts into our trough. We even made that hard by barging around and getting in the way, such was our excitement. They never seemed that fazed, though, and just worked around us.

Happy days indeed. Sheep, as a rule, like to be outdoors and hate being in sheds, but this spring, with that weather, it was just great.

As I mentioned, Peaches had lost her lambs, so her milk needed drying out. Because of this, she had to be kept away from the sheep nuts. I think that's why Peaches was taken out of the shed and put back in the field. It's not as bad as it sounds; there were two shelters so she could get out of the rain and plenty of fresh hay.

The removal of Peaches caused a bit of a stir. After all, she was our leader, our matriarch. How would she get on with no sheep to lead? Not only that, she was out there on her own, and sheep really don't like that. We wondered how the humans would solve that conundrum.

The solution – they chucked me out with her! What had I done to deserve that? No more sheep nuts for me. I was livid, depressed, every emotion you can think of. Trouble is, only Peaches could see that. Nevertheless, she was grateful for my company.

Thankfully, the weather did start to improve, so that helped a lot. Also, the grass was starting to grow; there's nothing quite like fresh, spring grass. We love hay, but spring grass is even better.

As the weeks passed, the other ewes came out to the field with their lambs. There was a bit of aggravation. Peaches seemed to take umbrage with some of the new mums and start a bit of headbutting (headbutting is how we sheep solve our differences). The humans often intervened to stop this. The male human even tried to pretend he was a tup and warn the aggressors off. That had us all in stitches. Luckily, the humans couldn't tell.

Anyway, one sunny morning it felt like it was time. There was a dog in the field and while I was generally OK with this particular dog, right now I wanted him out. He might have been a big dog, but I felt no fear as I chased him around the field. Annoyingly, he couldn't escape, but soon a human came to let him out. Right, it's time, I thought. Within minutes, I was popping out the first of my lambs.

The humans seemed quite shocked. Then it clicked – they had thought I was not pregnant. That's why they had chucked me out with Peaches. Just because I like to keep my figure, they must have thought I wasn't fat enough. Silly humans, I wasn't about to forget that.

It was a bit of a struggle, to be honest. The lack of sheep nuts meant we were all a bit thin. The humans tried to help, but they gave *me too many sheep nuts too quickly and I became quite ill. Also, my daughter Witchy had to be taken away for a while as she was quite poorly. But we all survived.*

Well, time went by during which we moved to a new place. It is fantastic. Not just two little paddocks but acres and acres of grass and custom-built field shelters we can all fit into. We were all grateful to our humans for that. Mind you, I hadn't forgotten 2017.

We went a year without lambs, which was fine. It gave us a bit of a rest, to be honest. For some reason the humans had brought in three new ewes, all pregnant. One of those lost her lambs and was put in with us. She'd had an operation and ended up with a huge stitched-up wound down one side. I have to confess, I was a bit mean to her. Even us sheep can be funny about new additions to the flock.

Anyway, time passed and that autumn we found ourselves in a field with a tup. Not all of us as some had been separated off. We tried to figure this out but couldn't really.

These tups were young whippersnappers with more energy than sense. The three new girls were straight in there all nestling in and coy. Me, I remembered 2017, so I hatched my plans.

The humans had built a new lambing facility, which was perfect. A good-sized shed, water on tap, plenty of ventilation and good sheltered areas. On top of that, we were not shut in; we had access to grass. Every day, we had a good ration of sheep nuts. We were ecstatic.

My plan was working. Would the humans suspect?

As the weeks passed, one by one my friends were giving birth to their lambs. Things went much more smoothly this year. The facilities were tip-top and the vet seemed to be able to get there quicker when there was a problem. It was great watching the young lambs bounding around the paddock, jumping over rocks, haring around the shed.

All through lambing, a human would check on us frequently. Even through the night they'd be out every two hours or so to check we were OK. They looked more and more knackered.

One by one, my friends gave birth, leaving only me. I was last again.

The days passed, and the humans came out to check on me every two hours. Meantime, I was still getting my sheep nuts. I enjoyed every

223

minute and wondered how long it would take before they clocked me.

Sadly, it wasn't long. The vet had come out to attend to some problem with a lamb, and then she came over to look at me. Nothing wrong with me, I thought. What's this about? The vet checked me all over, inside and out, and spoke to the humans. They looked at me with a look that told me they had figured it out.

I wasn't pregnant. My plan had worked. I'd had a lovely pampered spring with lots of attention and a steady supply of sheep nuts. I had kept my humans up at nights for days longer than they needed. My revenge was complete.

Luckily, the humans saw it as a great joke and there were no hard feelings. These days, they always give me back scratches – and they even know exactly which spot to scratch.

30

LAST WORD

My grandparents farmed a small farm on the west coast of Ireland. As seems common these days, the next generation (my father and his siblings) all moved away, mostly to Scotland and England to find a "better" way of life. But I think my family's history explains why, for me, farming has always been in my blood.

I have always felt a connection to the land. My father might have disagreed having witnessed my tenacious attempts to avoid helping him the garden. But that wasn't about the gardening.

Once I finally had my own patch, I was straight out reworking it. I was never a flower beds and tulips kind of gardener. All my gardens were modelled to become wildlife havens with a fruit and vegetable patch on the side. I have built a lot of ponds in my lifetime.

The reason for mentioning this is that I always had a dream of owning more land and keeping animals. After I left corporate life, I bought a house next to ancient woodland. I used to spend a lot of time in it. However, my small garden never felt quite big enough, especially when I started keeping hens.

I started looking around for something bigger. Not having a lot of money, the Home Counties, where I lived, were out of the question. I researched properties in Scotland (where I am from), Wales, Spain, France and even

Canada.

Then I met Nicole. It turned out that she had similar dreams and ancestry, her grandparents having owned a farm in their native Switzerland. We pooled our resources and bought a small smallholding in Somerset. It was only one and a quarter acres but, to us, it was huge.

I carried on with my software business, and Nicole built up a new gardening business from scratch. It was not long before we were both working full-time. On top of that, we started sorting out our patch. Nicole threw herself into creating new flower beds. I focused on the vegetable garden. We found ourselves working flat out.

However, we couldn't keep up with the grass, so we bought some sheep. That involved a lot of new fencing and equipment as well as a steep learning curve. I have never forgotten asking the vet to come out to do the sheep's annual vaccination only for him to talk us through the process, hand us the needle and tell us to get on with it. A bit different from the world of dogs and cats.

The irony here was that while the sheep ate the grass, in the summer months they couldn't keep up, so we had to get a tractor and topper to keep the grass under control. The tractor needed a shed, and so it continued.

We found ourselves working all hours, every day, and every penny we spent seemed to vanish into thin air.

That didn't stop us, though, and when a nearby field came up for sale, we bought that. More fencing, more sheep, plus we planted over a thousand trees.

The thing is, while it was hard work and proving more expensive than we had imagined, we loved the life itself. But we also needed to find a better balance. Aside from taking a lot of time and effort, the physical nature of smallholding life was taking its toll on us.

So, what did we do? We moved to an even bigger place. We did this mainly for the sheep. Our smallholding in Somerset could only really support around 10 sheep. We'd actually had to sell two of the first sheep we had bought. We have always kind of regretted that. We reasoned that if we were to continue lambing, we'd have to sell all the lambs. That kind of felt wrong.

The natural decision seemed to be to find a bigger place where we could support an influx of lambs as well as giving the sheep a good home.

We moved to Scotland buying a place with just over 27 acres. It was ideal for the sheep. However, it was not as well set up as we had thought. There was not really a decent lambing shed and there was no water system in the fields. We got to work.

We had not put our ewes in lamb, figuring that it would be too risky as we didn't know when we'd be moving. To compensate, we bought three ewes in lamb as soon as we moved.

We also took on pigs, which resulted in the construction of a pigpen. In fact, we found ourselves sorting out the post-move chaos, learning how to keep pigs, supervising home renovations (including much DIY) and lambing all at the same time.

Looking back, it was too much too quickly. It's all too easy to do, though, and that's part of the problem. In fact, during that first year, we also took on Highland cows. We loved those cows, but they were a step too far. Aside from taking up a lot of time and causing us problems (one of them was Houdini reborn), they proved very expensive. Within days of taking delivery, we had a letter from Animal Health asking for a TB test to be arranged. This would require a cow crush and holding area. For horned cattle, this was around £5,000 worth of equipment.

With both of us working, there was not enough time in the day to look after hens, sheep, cows and pigs, let alone ourselves. On top of that, all our money was going on farm equipment. As well as the cow crush, we had purchased haymaking machinery as well.

We had got the balance wrong. The smallholding life was costing us a fortune and wearing us out physically and mentally.

The relentlessness of it all had us yearning for a small flat, somewhere warm, where we could live out our life in peace.

It's all too easy to become disheartened. The books don't tell you about that. All that rosy nonsense about feeding a family from an acre. It is not that easy.

We had a conference and decided to scale back. The pigpen is now planted up as woodland. The cows and all their equipment went to a farm up the road. The buyers came from a family background of keeping cows, so they have gone to a good home. We have scaled the sheep back from 36 to 19. Our limit is 20.

Nicole has started to make products from the wool, and these are proving popular. The sheep are paying for themselves at long last.

I read an article in the papers recently about people who are skint but not poor. It kind of summed us up. We make enough to get by, but there are no luxuries anymore. Holidays are a thing of the past. A meal out is a real treat.

Now we have a better balance, we find the smallholding life more rewarding. We have a bit more time so we can have quality time with the sheep. And that is the key: balance. It is all too easy to take on too much, for example, by taking on more or new animals. Looking back, it would have made things easier had we given ourselves a bit more time before rushing in.

It might be easy to think, in retrospect, that had we thought things through we would never had made the leap. But the truth is, we treasure every experience. Yes, a bit more forewarning of what was involved, a bit more planning, a bit more consultation would have helped. We certainly might have avoided some large expenses.

But, all in all, the smallholding life has enriched our lives beyond measure.

ABOUT THE AUTHOR

Born in 1960, Adrian grew up in the city of Edinburgh, Scotland. Over time, he acquired a degree (BSc Hons) from Edinburgh University and migrated south, first to London and then the Home Counties. Starting out as a computer programmer, Adrian collected an MBA and rose up to a senior level of management within a global IT company. Adrian finally decided corporate life was not for him, so, in late 1999, he left said IT company and struck out on his own.

Now living in south-west Scotland with his wife, Nicole, 2 dogs, 19 sheep and 8 hens, Adrian has started writing and looks forward to trying his hand at fiction.

Check out Adrian and Nicole's website at www.auchenstroan.com for their latest news and products available to buy.

Printed in Great Britain
by Amazon